DATE			

Shaft Alignment
Handbook

MECHANICAL ENGINEERING

A Series of Textbooks and Reference Books

EDITORS

L. L. FAULKNER

Department of Mechanical Engineering
The Ohio State University
Columbus, Ohio

S. B. MENKES

Department of Mechanical Engineering
The City College of the
City University of New York
New York, New York

ADDITIONAL VOLUMES IN PREPARATION

Mechanical Engineering Software

Shaft Alignment Handbook

JOHN PIOTROWSKI

Manager, Air Supply Operations
General Electric Company
Cincinnati, Ohio

MARCEL DEKKER, INC.   New York and Basel

Library of Congress Cataloging-in-Publication Data

Piotrowski, John, [date]
    Shaft alignment handbook.

    (Mechanical engineering ; 46)
    Includes bibliographies and index.
    1. Machinery--Alignment. 2. Shafting. I. Title.
II. Series.
TJ177.5.P56 1986   621.8'23   85-29217
ISBN 0-8247-7432-9

MARCEL DEKKER, INC.
270 Madison Avenue, New York, New York 10016

Current printing (last digit):
10 9 8 7 6 5 4 3 2 1

PRINTED IN THE UNITED STATES OF AMERICA

To Bobbie Jo
for her patience and support

Preface

Many of the material conveniences taken for granted in today's
society have been made possible by the numerous rotating machinery systems located in every part of the world. The demand for
electricity, fossil fuels, chemicals, and transportation has increased dramatically in just the past few decades. The design and operation of this equipment has become considerably more complex in
comparison to the paddle wheels and reciprocating steam engines
that existed at the beginning of the Industrial Revolution. If
only Hero of Alexandria (a pioneer of mechanics in the third cen-
tury) could see the rows of powerful steam turbines lined up in
today's electric generating stations or the monstrous petrochem-
ical complexes of our era, requiring thousands of pieces of rotating equipment that consume millions of horsepower of energy.
All of this is indeed awesome; the ability to maintain peak performance in this machinery is even more so.

It makes good sense to want to keep every generator, motor,
pump, compressor, gear, and turbine operating for long periods
of time. Repair or replacement of this equipment is expensive

and the loss of revenue while machinery is down can spell the difference between continued prosperity or financial disaster. Keeping these machines running requires a thorough understanding of their design and operating envelope, careful attention during installation and overhauls, the faculty to predict imminent failures, and the expertise to modify existing hardware to extend its operating lifespan.

In the past ten years, easily half of the rotating equipment problems I have experienced had something to do with misaligned shafts. In discussing this with many other rotating equipment experts I've found that I'm not the only one who has noticed this to be true. Furthermore, misaligned machinery can be dangerous if not aligned correctly. I have seen a coupling burst apart on a 500 hp, 3600 rpm process pump that literally sheared a 10" pipe in half and landed 400 yards away from its original location. Keep in mind that rotor speeds up to 100,000 rpm and drivers pushing 60,000 hp are now commonplace.

How complicated could this shaft alignment process be? Can't one just set the machinery on its base and eyeball it in? Seems much ado about nothing, you might say. This book is meant to address these questions, and to show mechanics, foremen, technicians, and engineers how to properly align the shafts of two or more pieces of rotating equipment, thus minimizing vibration and reducing the wear that occurs in unnecessarily overloaded bearings and flexible couplings. Ultimately, responsibility for all this powerful machinery rests in the hands of each of these people and their managers.

The mechanics who actually measure the shaft positions, slide the equipment sideways, add or remove vertical shim packs, bolt the equipment down, and install the coupling are only a small part of the overall team effort that makes a precision alignment job successful. The maintenance or contractor foremen must know what steps are to be performed in sequence and insure each task is properly completed before continuing on to the next one. The engineers and technicians must be able to measure and calculate the movement of machinery from rest to running condition, specify the initial shaft positions, purchase the right kind of flexible coupling for the job, measure and understand vibration signatures of the machinery to know if a more precise alignment is required, inspect the foundation and baseplates for damage, and provide technical guidance to the foremen and the mechanics for each step of this process. Middle and upper management must insure that enough time is set aside to do the alignment job, supply the necessary funding for tools, hire the people to do the job, see that the proper training is furnished, provide the inspiration to want to do it right the first time, and, when it's due, give some recognition for a job well done.

A considerable amount of information has been compiled to help you understand what is involved in aligning this critical machinery. Many of the references cited in this book represent outstanding technical achievements in themselves and would provide the reader with further insight into this deceptively simple task. Generally, poor training, improper tooling, and lack of desire cause machinery to be misaligned. This book should help with the training. The rest is up to you.

John Piotrowski

Contents

Shaft Alignment
Handbook

1
The Need for Proper Alignment

Since the early 1900s, many technical articles have been pub-
lished on the subject of shaft alignment and its role in the field
of rotating equipment. Misaligned shafts have caused a tremendous amount of financial loss to every industry and are therefore of concern to every manager, engineer, supervisor, and
mechanic. The capacity to have all your rotating equipment well
aligned and running smoothly is directly related to your knowledge, ability, and desire to do it properly. It seems foolish to
install well-designed and constructed machinery only to watch
it be destroyed because no one wanted to spend six hours carefully aligning the shafts to allow the equipment to run for six
years instead of six months.

Tremendous advances have been made in just the past 20
years in diagnosing machinery health using vibration analysis
as a tool for pinpointing problems. Being able to discern each
of these maladies by analyzing vibration signatures and various
other data such as bearing temperatures and equipment performance data has been one of the fastest growing areas of technology.

Machinery problem diagnostic techniques have come a long way
from balancing nickles on bearing caps, vibrating light meters,
and the human touch. Many companies have based their entire
business on monitoring machinery and helping people locate their
equipment problems using FFT analyzers, spike energy meters,
and a host of other newly developed equipment aimed at understanding the dynamic response of turbomachinery.

Despite the rapid advances made in dynamic machinery analysis, advancement in obtaining accurate shaft alignment across
all industries has been sporadic at best. For many years, the
responsibility for aligning equipment has been left to millwrights
and mechanics. These people have not been adequately trained
in proper alignment techniques, and they have been subjected to
using inadequate tooling to measure the misalignment and move
the machinery. There always seems to be no interest on the part
of management to allow the personnel enough time to do a quality
job. Even in organizations where attention is paid to aligning the
critical, higher horsepower units, much of the smaller horsepower
units are ignored and poorly aligned, resulting in many premature
bearing or coupling failures.

Surprisingly, 99% of the turbomachinery operating today is
misaligned. This may sound like an extremely strong statement
but perfect alignment is nearly impossible to achieve. In fact, a
small amount of misalignment is really not that bad. Gear type
couplings and shafts with universal joint drives must have some
misalignment in order to keep lubrication at the points of power
transmission during rotation. So whenever we deal with alignment, it is important to know when to stop moving the machinery.
There comes a point where no beneficial returns can be made if
alignment tolerances have adequately been met.

To further complicate the issue, industrial rotating equipment
never wants to stay put. As the equipment begins to rotate, a
wide variety of factors contribute to moving the shafts somewhere
other than where they were when the units were sitting at rest.
Heat generated in the casings from bearings, lubrication systems,
movement of fluids, compression of gases, foundation movement
and settling, all have a tendency to move the equipment in every
conceivable direction. For shafts to run colinear at normal operating conditions, the equipment must be positioned correctly before the units are started to account for this movement. It is important to measure this movement where the equipment is located.
Do not take the manufacturer's word on calculated growth rates
or how the equipment moved on the test stand. Rarely are these
calculations correct because your equipment may be piped or

mounted differently from the ideal or test stand conditions. And don't pretend that you have the only machinery in existence that doesn't move around.

TYPES OF MISALIGNMENT

Shaft misalignment can occur in two basic ways: parallel and angular, as shown in Figure 1.1. Actual field conditions usually have a combination of both parallel and angular misalignment, so measuring the relationship of the shafts gets to be a little complicated in a three-dimensional world. This is especially true when you try to show this relationship on a two-dimensional piece of paper. I find it helpful at times to take a pencil in each hand and position them, based on the dial indicator readings, to reflect how the shafts of each unit are sitting.

HOW MISALIGNMENT IS DEFINED

To the person selecting a coupling for service, it is confusing how a coupling manufacturer specifies misalignment. Often misalignment is specified in *degrees* or radians (1 degree = 0.01745 radians) or *offset* in mils (1 mil = 0.001 inches). Sounds simple enough, but that's exactly where the confusion lies. Take for instance two different shaft arrangements, as shown in Figure 1.2.

For both conditions, the misalignment angle theta (θ) and the offset (distance X) is the same. However, condition no. 1 shows a pure parallel shaft relationship, whereas condition no. 2 shows a purely angular relationship. The vibration response of equipment subjected to condition no. 1 will be different and probably more severe compared to condition no. 2. Some coupling manufacturers will give maximum allowable misalignment capacities in degrees for angularity and mils for parallelism or may quote mils for both angularity and parallelism. The coupling manufacturer should be able to specify whether these misalignment limits are for a combination of angular and parallel or if they are strictly pure angular or pure parallel. It should also be specified what effect axial spacing discrepancies or dynamic axial movement will have on the coupling.

parallel

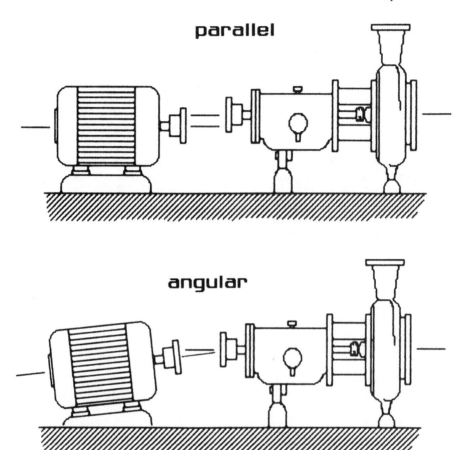

angular

FIGURE 1.1 How shafts can be misaligned.

Throughout this book misalignment conditions will be specified in mils offset occurring at each point of power transmission, with particular attention being paid to the maximum offset that occurs at either transmission point, as shown in Figure 1.3. Figure 1.4 shows the relationship between misalignment angles, distance between power transmission points, and
offset in mils.

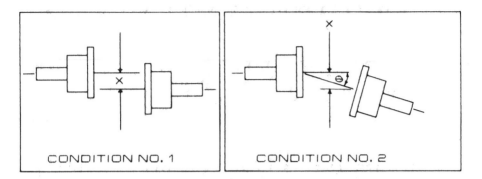

FIGURE 1.2 Specifying the amount of misalignment.

KEEPING RECORDS

It is helpful to keep data on the complete alignment process for each unit in your plant. These records should be kept in the maintenance file folder as a handy reference for future work done on each drive train. The machinery data card (shown in the Appendix of this book) can be used as a guide in planning your own record sheet.

HOW LONG SHOULD THE ALIGNMENT PROCESS TAKE?

If a mechanic performs an alignment job on a small pump for instance, once a month, and (1) knows how to take dial indicator readings; (2) knows how to calculate the necessary machinery moves; (3) has information from his or her engineers on thermal movement of the units; (4) has the proper tools at the job site; (5) does not have to fight against the pump piping if the pump has to be moved; (6) has a wide variety of precut shim stock; (7) has no coupling hub or shaft runout; (8) has no dirt, rust, or scale buildup under the feet; (9) has jackscrews installed on both units to lift and slide them sideways; (10) has shafts that rotate freely and no coupling pieces missing; (11) has correct

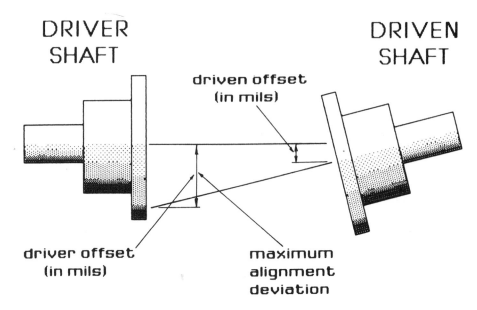

FIGURE 1.3 How misalignment is defined.

shaft-to-shaft distance, and no one to bother him or her, the
alignment should be completed with the coupling installed and the
coupling guard in place in about three to four hours. For people
who have never performed an alignment job, the last statement
may seem quite comical; but for people who have read this and
know what I mean, there is absolutely no humor in what was
written. I have *never* performed an alignment task and had
everything fall in place. If I ever do, I have definitely overlooked something. Measuring thermal movement alone can take
weeks on a multi-unit drive train. The heavier the equipment
is, the longer it takes to lift it or move it sideways.

There is a lot of time spent preparing for an alignment job.
Cleaning baseplates and the underside of the equipment feet, fab-
ricating dial indicator brackets, determining bracket sag, inspect-
ing the coupling, finding a soft foot, measuring thickness of shim
packs that are already installed, retapping foundation bolt holes,
gathering all the tools together, and spending some time and energy training personnel to do the job right are just a few of the

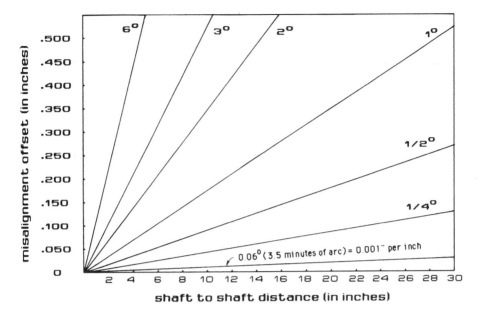

FIGURE 1.4 Alignment offset and angles based on power transmission distances.

things that have to be done before you start. Calculating the
proper moves needed to bring the shafts into alignment with computer or graphical alignment calculators can drastically reduce the
amount of time spent moving the machinery around compared to
trial-and-error methods. Achieving what I call "one-shot" alignment, that is, going straight from a rough alignment move to near
perfect alignment, happens about once in every ten tries. One
relatively large move and one "trim" move both sideways and vertically will usually achieve the desired results. If four or five
tries are needed to get the job done, something is wrong. Chapter 8 will explain where and how to look for many of these problems.

The intent of this book is to explain how to obtain accurate
shaft alignment to the contractor or maintenance mechanic, supervisor, and engineer whose work is in the field on a day-to-day
basis.

SUMMARY OF OVERALL ALIGNMENT JOB

Reference chapter	Step	Task
1, 3	1	Identify machinery to be aligned and relevant information on machine elements, coupling, etc. Use machinery data card (see Appendix) as reference.
2, 10	2	Inspect machinery foundation and baseplate for damage or problem areas. Plan necessary repairs or fixes during shutdown. Measure vibration levels and vibration signature of the equipment.
2, 7	3	Calculate the potential dynamic machinery movement or consult manufacturers for estimated movement values if new installation. Take actual field measurements for existing equipment using one or more of the techniques outlined in Chapter 7.
All	4	Train maintenance or contractor personnel on proper alignment measuring methods, machinery movement devices and procedures, checks for static movement of casings, obtaining cold alignment readings, use and care of dial indicators and micrometers, and allowable misalignment tolerances.
4, 7, 8	5	Purchase necessary tools and measuring devices. Fabricate alignment brackets, jackscrews, and lifting mechanisms. Measure bracket sag.
	6	Shut equipment down and take necessary precautions to prevent starting driver unit, and secure driven unit to prevent reverse rotation, etc.
4	7	Remove coupling guards and coupling spool piece if necessary. Measure the shaft and coupling hub runout on both units. Inspect coupling for any damage or worn parts. Keep track of the coupling parts and label them if necessary.

Reference chapter	Step	Task
		If equipment is going to be removed for servicing, make punch marks on casing and baseplate as a reference during reinstallation.
2, 8	8	Check for: "soft" foot, casing or frame warpage, excessive static piping or conduit forces, damaged or rusty shims, loose foundation bolts, etc. Correct any of the potential problem areas.
5	9	Take and record a set of alignment readings.
6, 7, 9	10	Calculate and perform the necessary axial, vertical, and horizontal moves. Re-check alignment, calculate, and move until equipment is within acceptable alignment tolerances. Torque foundation bolts to required values. Record final alignment readings.
	11	Install coupling and torque bolts to required values and install coupling guard. Check for rotational freedom of entire drive train if possible.
10	12	Operate unit at normal conditions checking and recording vibration levels, bearing temperatures, and other pertinent operating parameters.

2
Foundations, Baseplates, and Machine Casings

Rotating equipment alignment problems often result from problems in the foundation or machine casing. It is logical to conclude that the shaft alignment will change if there is a shift in the position of the foundation. This shifting can occur very slowly as the base soils begin to compress from the weight and vibration transmitted from the machinery above. It can also occur very rapidly from radiant or conductive heat transfer from the rotating equipment itself heating the soleplates, concrete, and attached structure. Piping strains on rotating equipment can be enormous and frequently cause drastic movement in the turbomachinery. These dynamic forces and the effect they have on rotating equipment can be measured very accurately in the field and is the subject of Chapter 7.

With the advent of computer technology and better design knowledge, foundations, structures, and machine casings can be rigorously designed and checked utilizing computer-aided design (CAD) and engineering (CAE) techniques before fabrication ever begins. The field of structural dynamics has provided the means to calculate structural mode shapes and system resonances of complex structures to insure that frequencies from the attached or adjacent machinery doesn't match the natural frequency of the structure itself. However, this newly applied technology can't easily remedy all the equipment already installed, and many of us are saddled with equipment sitting on poorly designed or constructed bases that are cracked or warped, or piping strains that have increased from the foundation settling over a period of time. It is unlikely that every foundation with a problem can be removed, redesigned, and installed, so it is important to understand how to deal with the variety of problems that can arise. This chapter will provide the reader with some techniques to check equipment in the field to determine if problems exist between the machinery and the foundation to which they are attached.

WHAT TO LOOK FOR

A complete visual inspection should be made at least once a year of all rotating equipment foundations, baseplates, piping, etc. A checklist for visual inspection is shown in Table 2.1. Many of these problems are quite obvious, as shown in Figures 2.1 and 2.2.

CHECKING THE FOUNDATION TO MACHINE CASING INTERFACE

Even with properly designed foundations, factors such as concrete shrinkage, thermal warpage, and settling of base soils can warp or buckle the foundation at the points of contact, with the machine casing causing the equipment to sit unevenly on its base. To determine if this problem exists, some relatively easy steps can be performed to determine if all the machinery feet are evenly loaded.

1. Mount dial indicators on the foundation near each corner of the machine element, as shown in Figure 2.3.

TABLE 2.1   Visual Inspection Checklist

Properly positioned piping hangars that carry the weight of the
piping

Piping expansion joints that move freely to accept thermal or hydraulic movement

Loose piping flange bolts

Cracked concrete bases or support columns

Cracks propagating at concrete joints

Water seeping between baseplate and concrete foundation that
could freeze and damage the structure

Loose foundation bolts

Shim packs that worked loose

Rusty shims

Loose or sheared dowel pins

Paint on shims

---

2. Tighten all the foundation bolts to their required torque value,
    with the shim packs installed under each foot, and zero all the
    dial indicators at each corner.
3. Starting at one corner of the machine element, loosen the
    foundation bolt at that corner and observe the reading on
    the dial indicator to determine if that foot lifts up.
4. Continue around the machine, loosening each corner bolt,
    and check each dial indicator where the foundation bolt has
    already been loosened.
5. If any foot has lifted more than 0.002 in., place a shim under
    that foot (or feet) equal to the amount of movement shown by
    the dial indicator.

It is possible to have a bowed condition along one side or
across adjacent corners, as shown in Figure 2.4 or Figure 2.5.

If the dial indicators are showing readings of 0.005 in. or
more, something is seriously wrong, and additional checks should
be made to determine the exact cause of the problem.

FIGURE 2.1 Loose shim pack and dowel pin.

CHECKING FOR EXCESSIVE STATIC PIPING
FORCES ON ROTATING EQUIPMENT

Since a majority of rotating equipment is used to transfer liquids
or gases, the connecting piping will undoubtedly have an effect
on the machinery and could potentially be another source of machinery movement due to thermal expansion of the piping, reac-
tionary forces from the movement of the liquid in the piping itself, static weight of the piping, or from piping that has not
been installed properly causing tension or compression at the
piping to machine interface. The forces that cause machinery
to move from improper installation of piping can be checked by
using dial indicators to monitor both the horizontal and vertical

FIGURE 2.2 Baseplate deteriorating from rust.

movement of the machine case. By placing indicators at each corner of the machine element, loosening all the foundation bolts,
and observing the amount of movement shown on the indicators,
any undesirable forces acting on the machine can be determined.
If more than 2 mils of movement is noticed, it may be possible to
reposition the other elements in the drive train without modify-
ing the piping to eliminate this problem. Chapters 6 and 7 will
explain how this can be accomplished.

This movement can also be checked with a shaft alignment
bracket attached to one shaft with dial indicators positioned at
the 12 o'clock and 3 o'clock position on the adjacent shaft, as
shown in Figure 2.6. Movement exceeding 2 mils on either dial
indicator is unacceptable after all the foundation bolts have been
loosened.

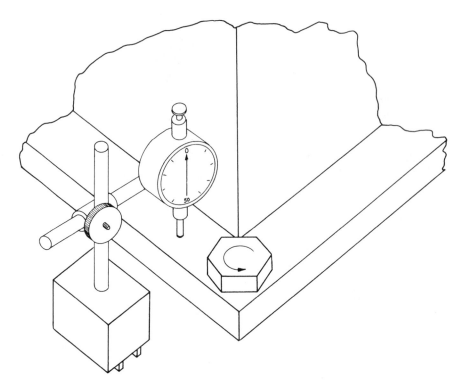

FIGURE 2.3 Checking for a "soft" foot or casing warp.

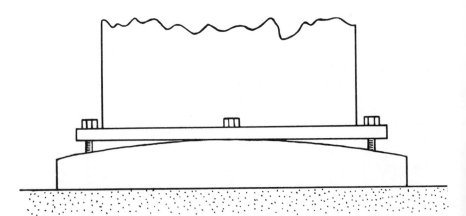

FIGURE 2.4 Bowed baseplate condition.

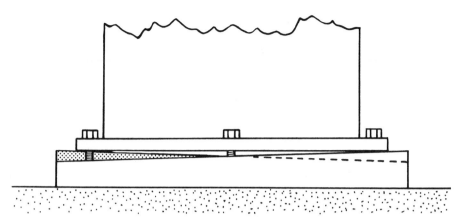

FIGURE 2.5 "Skewed" baseplate condition.

TIPS FOR DESIGNING GOOD FOUNDATIONS

1. Rotating equipment that will experience large amounts of thermal or dynamic movement from "cold" to "hot" should be spaced far enough apart to insure that the maximum allowable misalignment tolerance is not exceeded when the shafts are off-set in the "cold" position.

2. Insure that the natural frequency of the foundation/structure/soil system does not match any running machinery frequencies or harmonics (such as 0.43X, 1X, 2X, 3X, 4X, etc.), with the highest priority being placed on staying ±20% away from the operating speed of the machinery sitting on the foundation being considered. Also watch for potential problems where running speeds of any machinery nearby the proposed foundation might match the natural frequency of the system being installed. The design criteria for calculating natural frequencies of foundations and structures can be found in some of the reference literature at the end of this chapter.

3. In case the calculated natural frequency of the structure does not match the actual structure when built, design in some provisions for "tuning" the structure after erection has been completed, such as extension of the mat, "boots" around vertical support columns, attachments to adjacent foundations, etc.

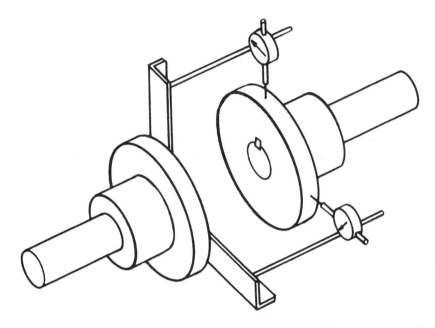

FIGURE 2.6 Using a shaft bracket to monitor machinery movement.

4. Design the foundation and structure to provide proper clearances for piping and maintenance work to be done on the machinery, and provisions for alignment of the machine elements.

5. Provide vibration joints or air gaps between the machinery foundation and the surrounding building structure to prevent transmission of vibration.

6. If possible, provide centrally located, fixed anchor points at both the inboard and outboard ends on each baseplate in a drive train to allow for lateral thermal plate expansion. Insure there is sufficient clearance on the casing foot bolt holes to allow for this expansion to occur without binding against the foot bolts themselves.

7. Minimize the height of the centerline of rotation from the baseplate.

8. Protect the foundation from any radiant heat generated from machinery, and steam or hot process piping by insulation or heat shields where possible.

TIPS ON INSTALLING FOUNDATIONS AND TURBOMACHINERY

1. Select a contractor having experience in installing rotating machinery baseplates and foundations or provide any necessary information to the contractor on compaction of base soils, amount and design of steel reinforcement, preparing concrete joints during construction, grouting methods, etc.

2. If the concrete for the entire foundation is not poured all at once, be sure to chip away the top 1/2 to 1 in. of concrete, remove debris, keep wet for several hours (or days if possible), allow surface to dry and immediately apply cement paste before continuing with an application of mortar (1 to 6 in.) and then the remainder of the concrete.

3. Use concrete vibrators to eliminate air pockets from forming during the pouring process but do not overvibrate causing the larger concrete particles to settle toward the bottom of the pour.

4. Check for baseplate distortion. Optical or laser tooling techniques can be used for this method. Refer to Chapter 7 for more information on how to use these tools. Mounting pads should be machined flat and not exceed 2 mils difference across all pads.

5. If the baseplate is distorted, stress relieve by oven baking or vibratory shakers.

6. Sandblast and coat the baseplate with inorganic zinc silicate per coating manufacturers' specifications to prevent corrosion and provide good bonding to grout.

7. Try not to use wedges to level the baseplate. This requires a two-step grouting procedure and may cause cracks to form in the grouting where the wedges were removed. If the wedges are left in after the grouting has been poured, there is a great likelihood that the grout will crack and separate where the wedges are located. Instead, weld 3/4 or 1 in. finely threaded nuts to the outside perimeter of the baseplate to use with jackscrews for precise leveling. Optics or lasers can be used to check level.

8. Grout one bulkhead section at a time. Apply grout through a 4 to 6-in. diameter hole centrally located in each section. Provide 1-in. diameter vent holes near the corners of each section. Refer to API specification 610 for additional grouting instructions. Allow a minimum of 48 hours cure time before setting rotating equipment onto base.

9. Install jackscrews for moving equipment in all three directions: vertically, laterally, and axially. If jackscrews will not be
used, provide sufficient clearance between baseplate and rotating
equipment for insertion of hydraulic jacks for lifting equipment for
shim installation. Refer to Chapter 8 on moving the machinery.

BIBLIOGRAPHY

Abel, L. W., D. C. Chang, and M. Lisnitzer. "The Design of
    Support Structures for Elevated Centrifugal Machinery."
    Proceedings, Sixth Annual Turbomachinery Symposium, Gas
    Turbine Labs, Texas A&M University, Dec. 1977, pp. 99-105.

Centrifugal Pumps for General Refinery Service. API Standard
    610, American Petroleum Institute, Wash., D.C., Mar. 1973.

Centrifugal Compressors for General Refinery Service. API
    Standard 617, American Petroleum Institute, Wash., D.C.
    (Oct. 1973).

Dodd, V. R. *Total Alignment*, Petroleum Publishing Co., Tulsa,
    Ok., 1975.

Essinger, Jack N. "A Closer Look at Turbomachinery Alignment."
    Hydrocarbon Processing (Sept. 1973).

Kramer, E. "Computations of Vibration of the Coupled System
    Machine-Foundation." Second International Conference, In-
    stitution of Mechanical Engineers and ASME, Churchill College,
    Cambridge, England, Sept., 1980, paper no. C300/80.

Massey, John R. "Installation of Large Rotating Equipment Sys-
    tems-A Contractors Comments." Proceedings, Fifth Turbo-
    machinery Symposium, Gas Turbine Labs, Texas A&M Uni-
    versity, College Station, Texas, Oct. 1976, pp.

Murray, M. G. "Better Pump Baseplates." *Hydrocarbon Pro-
    cessing* (Sept. 1973),

Murray, M. G. "Better Pump Grouting." *Hydrocarbon Process-
    ing* (Feb. 1974),

Newcomb, W. K. "Principles of Foundation Design for Engines
    and Compressors." ASME paper no. 50-OGP-5, April 1951.

Renfro, E. M. "Repair and Rehabilitation of Turbomachinery Foundations." Proceedings, Sixth Annual Turbomachinery Symposium, Gas Turbine Labs, Texas A&M University, Dec., 1977, pp. 107-112.

Renfro, E. M. "Five Years with Epoxy Grouts." Proceedings, Machinery Vibration Monitoring and Analysis Meeting, New Orleans, La., June, 1984.

Simmons, P. E. "Defining the Machine/Foundation Interface." Second International Conference, Institution of Mechanical Engineers and ASME, Churchill College, Cambridge, England, Sept. 1980, paper no. C252/80.

Sohre, J. S. "Foundations for High-Speed Machinery." ASME paper no. 62-WA-250, Sept., 1962.

Swiger, W. F. "On the Art of Designing Compressor Foundations." ASME paper no. 57-A-67, Nov., 1958.

Witmer, F. P. "How to Cut Vibration in Big Turbine Generator Foundations." *Power* (Nov. 1952),

3
Flexible Couplings

One of the most important components of any drive system is the coupling, the device connecting the rotating shafts. Since it is nearly impossible to maintain perfectly colinear centerlines of rotation between two or more shafts, couplings are designed to provide a certain degree of flexibility to allow for initial or running shaft misalignment. There is a wide assortment of flexible coupling designs, each available in a variety of sizes to suit specific service conditions.

The design engineer invariably asks, why are there so many types and is one type better than any other? Simply put, there is still no "perfect" way to connect rotating shafts. As you progress through this chapter, you will find that perhaps two or three different coupling types will fit the requirements for your drive system. One coupling being "better" than another is a relative term. If two or more coupling types satisfy the selection criterion and provide long, trouble free service, they are equal, not better. The ultimate challenge is for you to align shafts accurately, not find a coupling that can accept gross amounts of misalignment to compensate for your ineptitude.

It is important for the person selecting the coupling not to be confused by the term "allowable misalignment" in a coupling. The coupling manufacturers will often quote information on allowable misalignment for the coupling and not necessarily the equipment to which it is coupled. These tolerances seem to lull the user into a sense of complacency, leading one to believe that accurate shaft alignment is not necessary because "the coupling can take care of any misalignment."

The pursuit to connect two rotating shafts effectively dates back to the beginning of the industrial era, when leather straps and bushings or lengths of rope intertwined between pins were the medium used to compensate for shaft misalignment. As shaft speeds increased, coupling designs were continually refined to accept the new demands placed on them. As industrial competition became more severe, equipment downtime became a major concern and industry became increasingly more interested in coupling failures in an effort to prolong operating lifespans. Patents for diaphragm couplings date back to the 1890s but did not become widely used until just recently, as diaphragm design, material, and construction vastly improved. "O"-ring type seals and crowning of gear teeth in gear couplings came about during World War II. The awareness and concern for coupling and rotating machinery problems are reflected in the increase in technical information generated since the mid-1950s. Coupling designs will continually be refined in the coming years, with the ultimate goal of designing the "perfect" coupling.

THE ROLE OF THE FLEXIBLE COUPLING

Exactly what is a coupling supposed to do? If a perfect coupling were to exist, what would its design features include?

Allow limited amounts of parallel and angular misalignment
Transmit power
Insure no loss of lubricant in grease packed couplings despite
    misalignment
Install and disassemble easily
Accept torsional shock and dampen torsional vibration
Minimize lateral loads on bearings from misalignment

Allow for axial movement of shafts (end float), even under misaligned shaft conditions, without transferring thrust loads
   from one machine element to another
Stay rigidly attached to the shaft without damaging or fretting
   the shaft
Withstand temperatures from exposure to environment or from heat
   generated by friction in the coupling itself
Run under misaligned conditions (sometimes severe) when equipment is initially started to allow for equipment to eventually
   assume its running position
Provide failure warning and overload protection to prevent coupling from bursting or flying apart
Produce minimum unbalance forces
Have a minimal effect on changing system critical speed(s)

WHAT TO CONSIDER WHEN
SPECIFYING A COUPLING

Although some of the items listed below may not apply to your
specific design criteria when specifying a flexible coupling for
a rotating equipment drive system, it is a good idea to be aware
of all of these items when selecting the correct coupling for the
job.

Normal horsepower and speed
Maximum horsepower/torque being transmitted at maximum speed
   (often expressed as hp/rpm)
Misalignment capacity—parallel, angular, and combinations of both
   parallel and angular
Acceptance of the required amount of "cold" offset of the shafts
   without failure during start-up
Torsional flexibility
Service factor
Temperature range limits
Attachment to the shafts
Size and number of keyways
Type and amount of lubricant (if used)
Type and design of lubricant seals
Actual axial end float on rotors
Allowable axial float of shafts
Actual axial thermal growth or shrinkage of rotors

Type of environment to which coupling will be exposed
Likelihood of being subjected to radial or axial vibration from the
     equipment
Diameter of shafts and distance between shafts
Type of shaft ends (straight-bore, tapered, threaded, etc.)
Starting and running torque requirements
Cyclic or steady state running torques
Nature of failure—where it is likely to occur and what will happen
Noise and windage generated by the coupling
Cost and availability of spare parts
Lateral and axial resonances of the coupling
Coupling guard specifications for size, noise, and windage control
Installation procedure
Moments of inertia (WR2)
Heat generated from misalignment, windage, friction

TYPES OF FLEXIBLE COUPLINGS

The couplings found in this chapter show some of the commonly
used couplings in industry today but in no way reflect every
type, size, or manufacturer. The information presented for each
coupling concerning capacity, maximum speeds, shaft bore diameters, and shaft-to-shaft distances are general ranges and do
not reflect the maximum or minimum possible values available for
each coupling design.
     Misalignment capacities will not be given for a variety of
reasons.

1. Manufacturers of similar couplings do not agree or publish
    identical values for angular or parallel misalignment.
2. Manufacturers rarely specify if the maximum values for angu-
    lar misalignment and parallel misalignment are separate or a
    combination of the angular and parallel values stated.
3. It is the intent of this book to provide the reader with the
    ability to obtain alignment accuracies well within the limits of
    any flexible coupling design. Coupling manufacturers assume
    that the user will run the coupling within their stated maxi-
    mum misalignment values. If your rotating equipment or coup-
    ling has failed due to excessive misalignment, it is your fault.

This does not infer that all couplings accept the same maximum misalignment amounts or that these allowable values should not influence the selection of a coupling. Always consult with your coupling vendor about your specific coupling needs. If you are not getting the satisfaction you feel you need to select a coupling properly, consult a variety of manufacturers (or end users) to comment on design selection or problem identification and elimination.

Although there are a variety of coupling designs that accommodate fractional horsepower devices such as servomechanisms, this chapter will primarily show flexible couplings used on high horsepower, high-speed turbomachinery. However, to give the reader an idea of the design differences between the fractional and higher horsepower couplings, Figure 3.1 illustrates a few of the fractional horsepower designs.

Chain Coupling

The chain coupling shown in Figure 3.2 is basically two identical gear sprockets with hardened sprocket teeth connected by double width roller or silent type chain. Packed grease lubrication is primarily used with this type of construction, necessitating a sealed sprocket cover. A detachable pin or master link allows for removal of the chain. Clearances and flexing of the rollers and sprocket allow for misalignment and limited torsional flexibility.

Capacity: up to 1000 hp at 1800 rpm (roller), 3000 hp at 1800 rpm (silent)
Max. speed: up to 5000 rpm
Shaft bores: up to 8 in.
Shaft spacing: determined by chain width, generally 1/8 to 1/4 in.

Special Designs and Considerations

Wear generally occurs in sprocket teeth due to excessive misalignment or lack of lubrication. Torsional flexibility is limited by yielding of the chain.

(a)

FIGURE 3.1 Fractional horsepower coupling designs. Photos
courtesy of: (a and b) Metal Bellows Corp., Chatsworth, Ca;
(c) Guardian Industries, Michigan City, In.

ADVANTAGES

Easy to disassemble and reassemble
Fewer number of parts

DISADVANTAGES

Speed limited due to difficulties in maintaining balancing require-
    ments
Requires lubrication
Limited allowable axial displacement

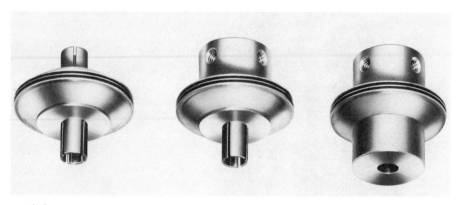

(b)

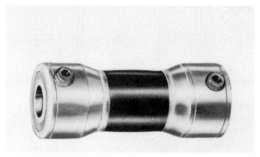

(c)

Diaphragm Coupling

Transmission of power occurs through two flexible metal diaphragms, each bolted to the outer rim of the shaft hubs and connected via a spacer tube as illustrated in Figure 3.3. Misalignment and axial displacement is accomplished by flexing of the diaphragm members:

Capacity: up to 30,000 hp
Max. speed: up to 30,000 rpm
Shaft bores: up to 7 in.
Shaft spacing: 2 to 200 in.

(a)

(b)

FIGURE 3.2 Chain coupling. Photos courtesy of: (a) Dodge/Reliance Electric, Greenville, S.C.; (b) Ramsey Products Corp., Charlotte, N.C.

FIGURE 3.3 Contoured diaphragm coupling. Photos courtesy of Koppers Co., Inc., Glen Arm, Md.

Special Designs and Considerations

Metal diaphragm couplings are a highly reliable drive component
when operated within their rated conditions. Exceeding the maximum allowable angular and/or parallel misalignment values or axial
spacing will eventually result in disc failure. Since the diaphragm
is, in effect, a spring, considerations must be given to the axial
spring rate and vibration characteristics to insure that the diaphragm coupling natural frequency does not match rotating
speeds or harmonics in the drive system.

ADVANTAGES

Excellent balance characteristics
No lubrication required
Low coupling weight and bending forces on shafts when operated
    within alignment limits
Acceptance of high temperature environment

DISADVANTAGES

Limited axial displacement and oscillation
Proper shaft spacing requirements generally more stringent than
    other coupling types
Excessive misalignment transmits high loads to shafting

Elastomeric Coupling

Figure 3.4 shows a wide variety of design variations employ an
elastomeric medium to transmit torque and accommodate misalignment. Couplings are torsionally soft to absorb high starting torques or shock loads.

Capacity: up to 67,000 hp/100 rpm (varies widely with design)
Max. speed: approx. 5000 rpm (varies widely with design)
Shaft bores: up to 30 in.
Shaft spacing: up to 100 in. (varies widely with design)

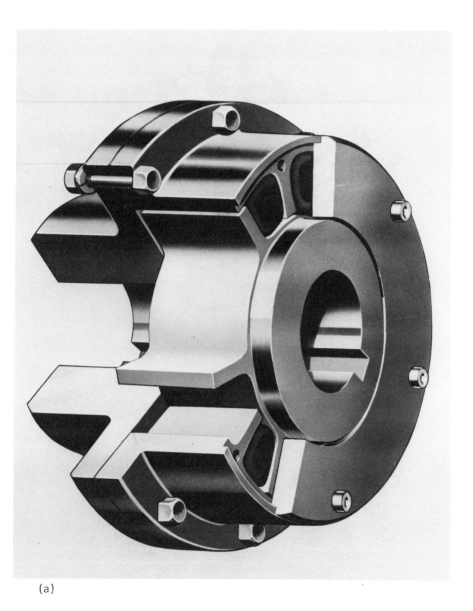

(a)

FIGURE 3.4 Various types of elastometric couplings. Photos
courtesy of: (a and b) Koppers Co., Inc., Glen Arm, Md.; (c)
Lovejoy, Downers Grove, Ill.; (d, e, and f) Dodge/Reliance Electric, Greenville, S.C.; (g) Falk Corporation, Milwaukee, Wis.;
(h) American Vulkan Corp., Winter Haven, Fl.; (i) T. B. Wood's
Sons Co., Chambersburg, Pa.

(b)

(c)

FIGURE 3.4 (continued)

(d)

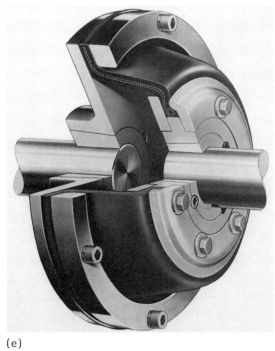

(e)

(f)

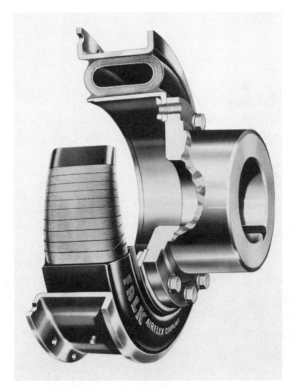

(g)

FIGURE 3.4 (continued)

(h)

(i)

Special Designs and Considerations

A considerable amount of inventiveness and ingenuity has been applied to this type of coupling design through the years as evidenced by the large array of design variations. The elastomeric medium is generally natural or synthetic rubber, urethane, nylon, teflon, or oil-impregnated bronze. Since the elastomer is markedly softer than the hubs and solid-driving elements (wedges, pins, jaws, etc.), wear is minimal and replacement of the elastomer itself is all that is usually needed for periodic servicing.

ADVANTAGES

Minimal wear in coupling
Acts as vibration damper and isolator
Acts as electrical shaft current insulator in some designs
Torsionally "soft"
Accepts some axial movement and dampens axial vibration
No lubrication required

DISADVANTAGES

Speed limited due to distortion of elastomer from high centrifugal
    forces, causing imbalance
Deterioration of elastomer possible from temperature, oxidation of
    rubber, corrosive attack from undesirable environment
Potential safety hazard if elastomeric member releases from drive
    elements
Some designs may cause undesirable axial forces
Heat generated from cyclic flexing of elastomer

Flexible Disc Coupling

The flexible disc coupling illustrated in Figure 3.5 is very similar in design principles to the diaphragm coupling with the exception that multiple, thinner discs or a noncircular flexing member is used as the flexing element instead of circular, contoured diaphragm elements.

Capacity: up to 65,000 hp/100 rpm
Max. speed: up to 30,000 rpm
Shaft bores: to 12 in.
Shaft spacing: to 200 in.

Special Designs and Considerations

It is important to note that two disc packs (or diaphragms) are needed to accommodate parallel misalignment, whereas a single disc can only handle pure angular misalignment. Convolutions in the discs provide a linear stiffness versus deflection characteristics as opposed to flat disc profiles. Once again, coupling axial resonance information must be known to prevent problems where a match may occur with machinery running speeds, higher order harmonics, or subsynchronous forcing mechanisms (oil whirl, looseness of bearing housings, clearance induced whirls, etc.).

ADVANTAGES AND DISADVANTAGES

Same as diaphragm couplings

Flexible Link Coupling

The flexible link coupling shown in Figure 3.6 utilizes a series of cross-laced, metallic links, with one end of each link attached to a disc mounted on the driven shaft and the other end of each link attached to a disc mounted on the driver shaft. The links are matched in pairs so that when one is in tension, the other is in compression. Misalignment and axial displacement is accomplished by a flexing action in the series of cross links.

Capacity: up to 1100 hp/100 rpm
Max. speed: to 1800 rpm
Shaft bores: up to 20 in.
Shaft spacing: close coupled or 100 mm spacer with certain designs

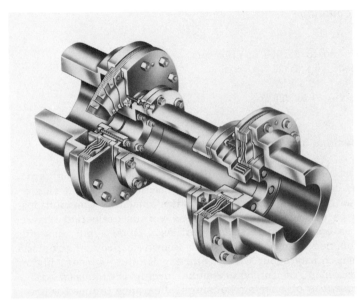

(a)

(b)

FIGURE 3.5 Flexible disc coupling designs. Photos courtesy of:
(a) Zurn Industries, Inc., Erie, Pa.; (b) Dana-Industrial Power
Transmission Division, San Marcos, Tx.; (c) Coupling Corp. of
America, York, Pa.; (d) Schmidt Couplings, Inc., Cincinnati,
Oh.

(c)

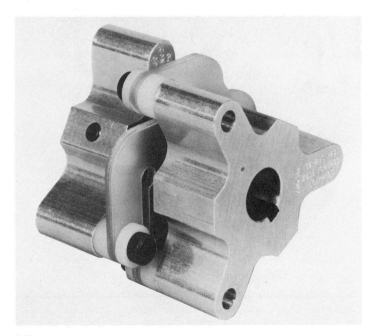

(d)

FIGURE 3.6 Flexible link coupling. Photo courtesy of Eaton Corp.,
Airflex Division, Cleveland, Oh., under license from Dr. Ing.
Geislinger & Co., Salzburg, Austria.

Special Designs and Considerations

An axial "fixation" device can be installed to prevent any axial movement if desired. Different designs can accommodate unidirectional or bi-directional rotation.

ADVANTAGES

No lubrication required

DISADVANTAGES

Limited axial movement
Limited misalignment capabilities

Gear Coupling

The gear coupling illustrated in Figure 3.7 consists of two hubs with external gear teeth that are attached to the shafts. A hub cover or sleeve with internal gear teeth engages with the shaft hubs to provide the transmission of power. Gear tooth clearances and tooth profiles allow misalignment between shafts. Lubrication of the gear teeth is required, and various designs allow for grease or oil as the lubricant.

Capacity: up to 70,000 hp
Max. speed: up to 50,000 rpm
Shaft bores: up to 30 in.
Shaft spacing: up to 200 in.

Special Designs and Considerations

A considerable amount of attention is paid to the form of the tooth itself (see Figure 3.8) and the tooth "profile" has progressively evolved through the years to provide minimum wear to the mating surfaces of the internal and external gear sets.

To provide good balance characteristics, the tip of the external gear tooth is curved and tightly fits into the mating internal gear hub cover. If the fit is too tight, the coupling will be unable to accept misalignment without damaging the coupling or the

FIGURE 3.7 Gear type couplings. Photo courtesy of Dodge/Reliance Electric, Greenville, S.C.

rotating equipment. If it is too loose, the excessive clearance will cause a condition of imbalance. Obtaining a good fit can become very tricky when the coupling hubs have been thermally or hydraulically expanded and shrunken onto a shaft where an increase

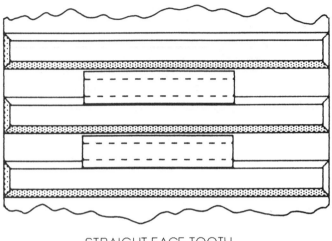

STRAIGHT FACE TOOTH

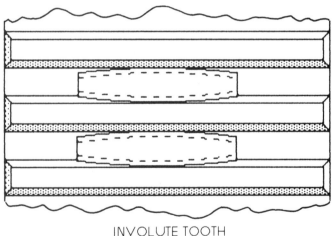

INVOLUTE TOOTH

FIGURE 3.8 Gear coupling tooth profiles.

in diameter of the external gears will occur. As a rule of thumb,
1 mil per inch of external gear tooth diameter can be used as the
clearance.

The amount of misalignment in a gear coupling directly affects the wear that will occur in the mating gear teeth. To better

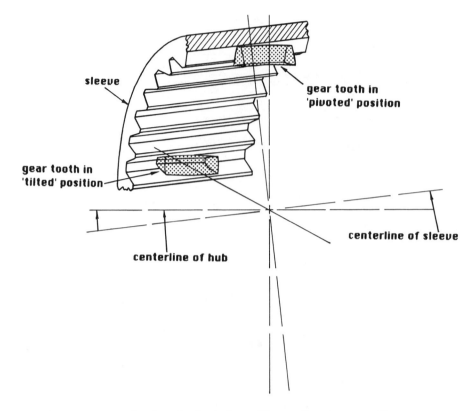

FIGURE 3.9 Tilted and pivoted positions of a gear coupling.

understand the motion of a gear coupling, Figure 3.9 shows
the two basic positions gear teeth will take in the coupling
sleeve.

At a certain point during the rotation of the shaft, the gear
tooth is in a *tilted* position and will completely reverse its tilt
angle 180 degrees from that point relative to the coupling sleeve.
Ninety degrees from the tilted position, the gear tooth now as-
sumes a *pivoted* position, which also reverses 180 degrees. The
gear tooth forces are at their maximum when in the tilted position as they supply the rotational transmission of power. As
the misalignment increases, fewer teeth will bear the load. De-
pending on the relative position of the two shafts, each set of

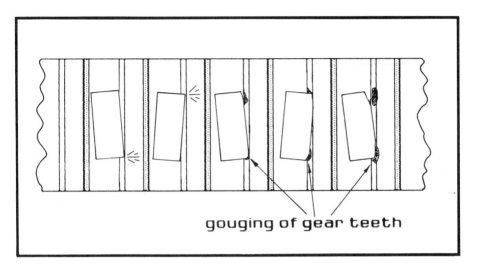

gouging of gear teeth

FIGURE 3.10 Gouging in gear teeth.

gear teeth on each hub may have their maximum tilt and pivot
points at different positions with respect to a fixed angular reference location.

Under excessive misalignment conditions, the load will be
carried by the ends of the gear tooth flank and eventually cause
gouging of the internal gear teeth and knife-edging of the external gear, since the compressive stresses are extremely high.
This forces any lubricating film out, allowing metal-to-metal contact to occur, as shown in Figure 3.10.

Another peculiar wear pattern evolves when the gear tooth
sliding velocity falls into the 5 to 8 inches per second range and
lubrication between the gear teeth diminishes. This type of wear
is known as *worm-tracking*, where gouges occur generally from
the base to the tip of the tooth flank as shown in Figure 3.11.
The formation of this type of wear pattern will occur when little
or no lubrication takes place at the points of contact among teeth,
and the metal-to-metal contact fuse welds a small portion of the
tooth flanks. As rotation continues, cracks begin to form at the
outer edges of the weld and eventually propagate until the two
welded pieces separate entirely from the mating external and internal gear teeth.

FIGURE 3.11 Tooth flank with "worm-track" wear pattern.

ADVANTAGES

Allows freedom of axial movement
Capable of high speeds
Low overhung weight
Good balance characteristics with proper fits and curved tooth
    tip profile
Long history of successful applications

DISADVANTAGES

Requires lubrication
Temperature limitation due to lubricant
Difficult to calculate reaction forces and moments of turbomachin-
    ery rotors when using these couplings since the values for
    the coefficient of friction between the gear teeth vary con-
    siderably

Leaf Spring Coupling

This coupling (Fig. 3.12) employs a series of radially positioned
sets of leaf springs attached to an outer drive member and indexed

into axial grooves in the inner drive member. The chamber
around each spring set is filled with oil. When the spring pack
is deflected, damping occurs as the oil flows from one side of
the spring pack to the other.

Capacity: up to 15,000 hp/100 rpm
Max. speed: 3600 rpm
Shaft bores: up to 12 in.
Shaft spacing: up to 40 in.

Special Designs and Considerations

The leaf spring coupling is designed primarily for diesel and reciprocating machines. It is capable of transmitting shock torque
values substantially higher than other couplings until springs
reach their maximum allowable angular movement where the radial
stiffness increases substantially. Various spring stiffnesses can
be installed in each size coupling to match the torsional requirements to the drive system properly.

ADVANTAGES

Torsionally soft with good damping characteristics
Freedom of axial shaft movement

DISADVANTAGES

Requires lubricant for damping
Temperature limitations due to lubricant
Torsional characteristics change drastically with loss of oil

Metallic Grid Coupling

Metallic grid couplings (Fig. 3.13) consist of two hubs with axial
grooves on the outer diameter of the hub where a continuous
S-shaped grid meshes into the grooves. Misalignment and axial

FIGURE 3.12 Leaf spring coupling. Photos courtesy of Eaton
Corp., Airflex Division, Cleveland, Ohio.

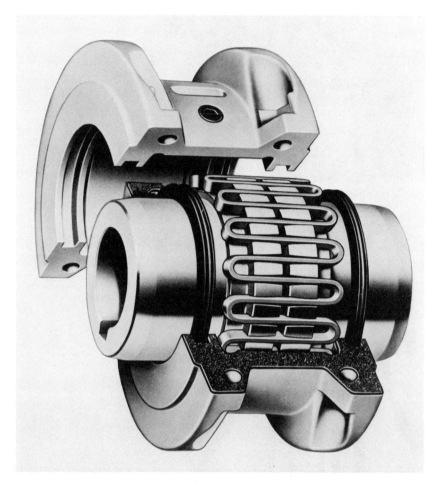

FIGURE 3.13 Metallic grid coupling. Photo courtesy of Falk
Corp., Milwaukee, Wi.

movement is achieved by flexing and sliding of the grid member
in specially tapered hub teeth.

Capacity: up to 70,000 hp/100 rpm
Max. speed: to 6000 rpm
Shaft bores: to 20 in.
Shaft spacing: to 12 in.

Special Designs and Considerations

The grid is fabricated from hardened, high strength steel. Close coupled hubs with a removable spacer are available. This coupling is used for vibration study in Chapter 10.

ADVANTAGES

Easy to assemble and disassemble
Long history of successful applications
Torsionally soft

DISADVANTAGES

Requires lubrication
Temperature limited
Speed limited

Pin Drive Coupling

A series of metal pins with leaf springs are placed near the outer diameter where they engage into a series of holes bored into both shaft hubs. (See Fig. 3.14.) Some pin designs consist of a pack of flat springs with cylindrical keepers at each end that act as the flexing element in the coupling design. The spring sets can swivel in the pin connection to allow movement across the width of the spring set.

Capacity: up to 3800 hp at 100 rpm
Max. speed: to 4000 rpm
Shaft bores: to 13 in.
Shaft spacing: close coupled (1/8 to 1/2 in.)

Special Designs and Considerations

Drive pins can be fabricated to accommodate various torsional flexibility requirements and are indexed into oil-impregnated bronze bushings in the coupling hubs.

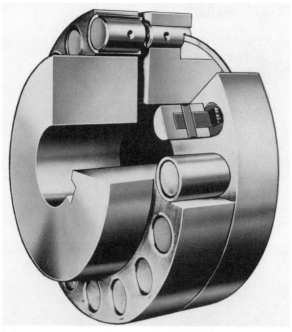

(a)

FIGURE 3.14 Pin drive coupling. Photos courtesy of: (a)
Dodge/Reliance Electric, Greenville, S.C.; (b) David Brown
Gear Industries, Agincourt, Ontario.

ADVANTAGES

Can accommodate up to 1/2 in. of axial displacement
No lubrication required

DISADVANTAGES

Limited misalignment capability

COUPLING LUBRICATION

There are basically two methods used to lubricate couplings: single charge and continuous feed. Greases are generally used in

(b)

single charge lubricated couplings and the type is generally specified by the coupling manufacturer.

Problems that can occur in greased packed couplings are:

1. Loss of lubricant from leakage at lube seals, shaft keyways,
   mating flange faces, or lubricant filler plugs
2. Excessive heat generated in the coupling from an insufficient
   amount of lubricant, excessive misalignment, or poor heat
   dissipation inside the coupling shroud that reduces viscosity
   and accelerates oxidation
3. Improper lubricant
4. Centrifugal forces generated in the coupling can be high
   enough to separate greases into oils and soaps

Since soaps have a higher specific gravity than oil, it will
eventually collect where the force is the highest (namely where
the gear teeth are located) causing a buildup of sludge.

Periodically inspect the inside of the coupling guard and directly under the coupling to see if any leakage is occurring. If

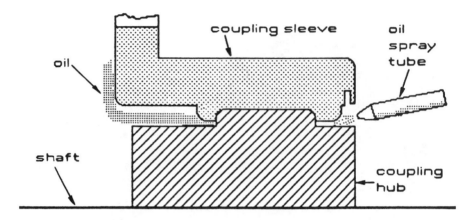

FIGURE 3.15 Continuous oil feed systems for a gear coupling.

so, do not continue to add more grease since the oil usually leaks out and the soaps continue to build up. Thoroughly clean the coupling, replace the seals and gaskets, and replenish with the correct kind and amount of lubricant.

Continuous feed lubrication systems generally use the same lube oil as the bearings, and spray tubes are positioned to inject a directed stream of oil into the coupling as shown in Figure 3.15.

In addition to supplying lubricant to the coupling, a continuous supply of oil acts as an excellent heat transfer agent, maintaining a relatively stable temperature in the coupling. However, contaminants in the oil, particularly water (which often condenses in lube oil tanks), or corrosive process gases carried over from the inboard oil seals on compressors can damage the coupling in time. Stainless steel lube oil piping, condensate and particulate matter removal with lube oil centrifuges, 5 to 10 micron filters, and entrained gas venting systems will alleviate many of these problems.

COUPLING INSTALLATION

Once a flexible coupling has been selected for a specific service, the next important step is proper installation. It is quite easy to destroy an expensive coupling assembly due to sloppy shaft fits, incorrect key dimensions, improperly measured shaft diameters, and so on. After the coupling has been uncrated the following steps should be performed before installation is even attempted.

1. Insure that the correct type of coupling was ordered and all the parts are with it (bolts, spacer spool, hubs, cover, gaskets, etc.).
2. Physically measure all the dimensions against the coupling drawing and parts listings, paying particular attention to coupling hub bores, keyway dimensions, and spool length.
3. Measure the shafts where the coupling is going to be installed (i.e., outside diameters, tapers, keyways, etc.).
4. If possible, assemble the entire coupling before it is placed on the shaft, checking for proper gear tooth clearances, elastomeric member fits, bolt hole diameter fits, and clearances.

COUPLING HUB ATTACHMENT METHODS

There is a variety of methods employed to attach the coupling hubs to a shaft, each one having its advantages and disadvantages. Recommended guidelines for installing these various shaft-to-coupling hub fits are outlined below and should be followed to insure a proper fit to prevent slippage or unwanted shaft fretting. Shaft fretting occurs when a coupling hub is loose on its shaft and the oscillatory rocking motions of the hub cause pitting on the mating surfaces of the shaft and the coupling hub.

Types of Coupling Hub-to-Shaft Fits

Straight bore, sliding clearance with keyway(s)
Straight bore, interference fit with keyway(s)
Splined shaft with end lock nut
Tapered bore, inteference fit with keyway(s)
Keyless taper bore
Locking taper cone

TABLE 3.1 Key and Keyway Sizes for Various Shaft Diameters[a]

| Nominal shaft diameter | | Nominal key size | | | Nominal keyset depth | |
| | | | Height, H | | H/2 | |
Over	To (incl.)	Width, W	Square	Rectangular	Square	Rectangular
5/16	7/16	3/32	3/32	.....	3/64	.....
7/16	9/16	1/8	1/8	3/32	1/16	3/64
9/16	7/8	3/16	3/16	1/8	3/32	1/16
7/8	1 1/4	1/4	1/4	3/16	1/8	3/32
1 1/4	1 3/8	5/16	5/16	1/4	5/32	1/8
1 3/8	1 3/4	3/8	3/8	1/4	3/16	1/8
1 3/4	2 1/4	1/2	1/2	3/8	1/4	3/16
2 1/4	2 3/4	5/8	5/8	7/16	5/16	7/32
2 3/4	3 1/4	3/4	3/4	1/2	3/8	1/4
3 1/4	3 3/4	7/8	7/8	5/8	7/16	5/16
3 3/4	4 1/2	1	1	3/4	1/2	3/8
4 1/2	5 1/2	1 1/4	1 1/4	7/8	5/8	7/16
5 1/2	6 1/2	1 1/2	1 1/2	1	3/4	1/2
6 1/2	7 1/2	1 3/4	1 3/4	1 1/2[b]	7/8	3/4
7 1/2	9	2	2	1 1/2	1	3/4
9	11	2 1/2	2 1/2	1 3/4	1 1/4	7/8

[a]All dimensions are given in inches. Square keys preferred for shaft dimensions above heavy line; rectangular keys, below.
[b]Some key standards show 1 1/4 inches; preferred height is 1 1/2 inches.
Source: Machinery's Handbook, 21st ed., Industrial Press, New York, 1980.

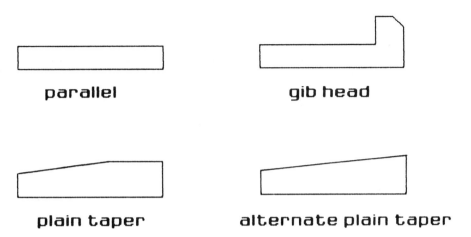

parallel   gib head

plain taper   alternate plain taper

FIGURE 3.16 Types of keys.

Keys and Keyways

A large percentage of shafts employ one or more keys to prevent
the coupling hub from rotating on the shaft as the rotational
force is applied. The American National Standards Institute
(ANSI) has set up design guidelines for proper shaft sizes to
key sizes, and these are shown in Table 3.1.

Types of Keys

Figure 3.16 shows the basic key designs used in industry. There
are three different classes of key fits:

Class 1: side and top clearance, relatively loose fit, only applies
     to parallel keys (see Table 3.2).
Class 2: minimum to possible interference, relatively tight fit
     (see Table 3.2).
Class 3: interference with degree of interference not standard-
     ized. Use Class 2 fits in Table 3.2 as a guideline.

TABLE 3.2 Key Fits[a]

Type of key	Key width		Side fit			Top and bottom fit			
			Width tolerance			Depth tolerance			
	Over	To (incl.)	Key	Key-seat	Fit range	Key	Shaft keyseat	Hub keyseat	Fit range
Class I Fit for Parallel Keys									
	.....	1/2	+0.000 -0.002	+0.002 -0.000	0.004 CL 0.000	+0.000 -0.002	+0.000 -0.015	+0.010 -0.000	0.032 CL 0.005 CL
	1/2	3/4	+0.000 -0.002	+0.003 -0.000	0.005 CL 0.000	+0.000 -0.002	+0.000 -0.015	+0.010 -0.000	0.032 CL 0.005 CL
	3/4	1	+0.000 -0.003	+0.003 -0.000	0.006 CL 0.000	+0.000 -0.003	+0.000 -0.015	+0.010 -0.000	0.033 CL 0.005 CL
Square	1	1 1/2	+0.000 -0.003	+0.004 -0.000	0.007 CL 0.000	+0.000 -0.003	+0.000 -0.015	+0.010 -0.000	0.033 CL 0.005 CL
	1 1/2	2 1/2	+0.000 -0.004	+0.004 -0.000	0.008 CL 0.000	+0.000 -0.004	+0.000 -0.015	+0.010 -0.000	0.034 CL 0.005 CL
	2 1/2	3 1/2	+0.000 -0.006	+0.004 -0.000	0.010 CL 0.000	+0.000 -0.006	+0.000 -0.015	+0.010 -0.000	0.036 CL 0.005 CL
	.....	1/2	+0.000 -0.003	+0.002 -0.000	0.005 CL 0.000	+0.000 -0.003	+0.000 -0.015	+0.010 -0.000	0.033 CL 0.005 CL

Rectangular

Size								
1/2	3/4	+0.000 / -0.003	+0.003 / -0.000	0.006 CL / 0.000	+0.000 / -0.003	+0.000 / -0.015	+0.010 / -0.000	0.033 CL / 0.005 CL
3/4	1	+0.000 / -0.004	+0.003 / -0.000	0.007 CL / 0.000	+0.000 / -0.004	+0.000 / -0.015	+0.010 / -0.000	0.034 CL / 0.005 CL
1	1 1/2	+0.000 / -0.004	+0.004 / -0.000	0.008 CL / 0.000	+0.000 / -0.004	+0.000 / -0.015	+0.010 / -0.000	0.034 CL / 0.005 CL
1 1/2	3	+0.000 / -0.005	+0.004 / -0.000	0.009 CL / 0.000	+0.000 / -0.005	+0.000 / -0.015	+0.010 / -0.000	0.035 CL / 0.005 CL
3	4	+0.000 / -0.006	+0.004 / -0.000	0.010 CL / 0.000	+0.000 / -0.006	+0.000 / -0.015	+0.010 / -0.000	0.036 CL / 0.005 CL
4	6	+0.000 / -0.008	+0.004 / -0.000	0.012 CL / 0.000	+0.000 / -0.008	+0.000 / -0.015	+0.010 / -0.000	0.038 CL / 0.005 CL
6	7	+0.000 / -0.013	+0.004 / -0.000	0.017 CL / 0.000	+0.000 / -0.013	+0.000 / -0.015	+0.010 / -0.000	0.043 CL / 0.005 CL

Class 2 Fit for Parallel and Taper Keys

Parallel Square

Size								
.....	1 1/4	+0.001 / -0.000	+0.002 / -0.000	0.002 CL / 0.001 INT	+0.001 / -0.000	+0.000 / -0.015	+0.010 / -0.000	0.030 CL / 0.004 CL
1 1/4	3	+0.002 / -0.000	+0.002 / -0.000	0.002 CL / 0.002 INT	+0.002 / -0.000	+0.000 / -0.015	+0.010 / -0.000	0.030 CL / 0.003 CL
3	3 1/2	+0.003 / -0.000	+0.002 / -0.000	0.002 CL / 0.003 INT	+0.003 / -0.000	+0.000 / -0.015	+0.010 / -0.000	0.030 CL / 0.002 CL

TABLE 3.2 (continued)

Type of key	Key width Over	Key width To (incl.)	Side fit — Width tolerance Key	Side fit — Width tolerance Key-seat	Side fit — Fit range	Top and bottom fit — Depth tolerance Key	Top and bottom fit — Depth tolerance Shaft keyseat	Top and bottom fit — Depth tolerance Hub keyseat	Top and bottom fit — Fit range
Parallel Rectangular	.....	1 1/4	+0.001 / -0.000	+0.002 / -0.000	0.002 CL / 0.001 INT	+0.005 / -0.015	+0.000 / -0.015	+0.010 / -0.000	0.035 CL / 0.000 CL
	1 1/4	3	+0.002 / -0.000	+0.002 / -0.000	0.002 CL / 0.002 INT	+0.005 / -0.005	+0.000 / -0.015	+0.010 / -0.000	0.035 CL / 0.000 CL
	3	7	+0.003 / -0.000	+0.002 / -0.000	0.002 CL / 0.003 INT	+0.005 / -0.005	+0.000 / -0.015	+0.010 / -0.000	0.035 CL / 0.000 CL
Taper	.....	1 1/4	+0.001 / -0.000	+0.002 / -0.000	0.002 CL / 0.001 INT	+0.005 / -0.000	+0.000 / -0.015	+0.010 / -0.000	0.005 CL / 0.025 INT
	1 1/4	3	+0.002 / -0.000	+0.002 / -0.000	0.002 CL / 0.002 INT	+0.005 / -0.000	+0.000 / -0.015	+0.010 / -0.000	0.005 CL / 0.025 INT
	3		+0.003 / -0.000	+0.002 / -0.000	0.002 CL / 0.003 INT	+0.005 / -0.000	+0.000 / -0.015	+0.010 / -0.000	0.005 CL / 0.025 INT

[a] All dimensions are given in inches.

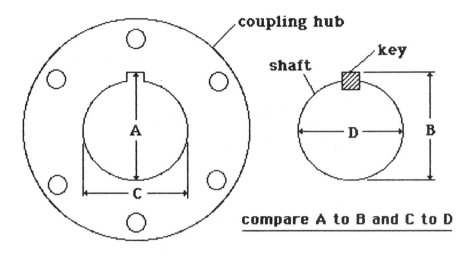

FIGURE 3.17 Measuring a coupling hub for proper key fits.

It is important to measure how the key will fit as it is to measure shaft diameter (as shown in Figure 3.17).

Shafts with two keyways can present another type of problem
from improper machining of the coupling hub or shaft keyways,
as illustrated in Figure 3.18.

If the offset gap is larger than 10% of the total key width, it
is recommended that the shaft or coupling hub be reworked for
improved indexing of the keys in their mating keyways.

Straight Bore—Sliding Clearance with Keyways

This method of shaft and coupling hub fitup is used extensively
in industry and provides the easiest and quickest installation of
the coupling hub. However shaft fretting is likely to occur with
this sort of arrangement since there is a certain amount of clearance (0.0005 to 0.001 in. generally) between the coupling hub and
the shaft. To prevent the coupling hub from sliding axially along
the shaft, set screws are usually locked against the key, as shown
in Figure 3.19. Use Table 3.3 for matching the correct size set
screw with the key size.

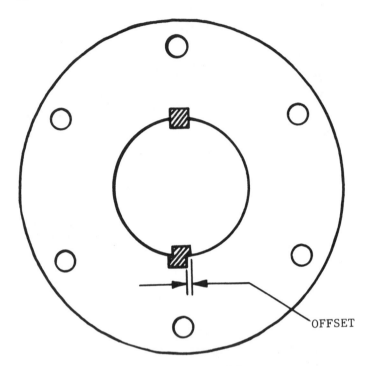

FIGURE 3.18 Improperly broached keyway.

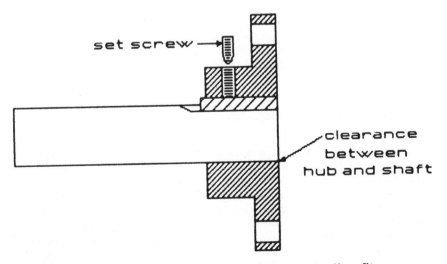

FIGURE 3.19 Straight bore with interference coupling fitup.

TABLE 3.3 Set Screw—Key Size Directory

Nom. shaft diam.		Nom. key width	Set screw diam.
Over	To (incl.)		
5/16	7/16	3/32	No. 10
7/16	9/16	1/8	No. 10
9/16	7/8	3/16	1/4
7/8	1 1/4	1/4	5/16
1 1/4	1 3/8	5/16	3/8
1 3/8	1 3/4	3/8	3/8
1 3/4	2 1/4	1/2	1/2
2 1/4	2 3/4	5/8	1/2
2 3/4	3 1/4	3/4	5/8
3 1/4	3 3/4	7/8	3/4
3 3/4	4 1/2	1	3/4
4 1/2	5 1/2	1 1/4	7/8
5 1/2	6 1/2	1 1/2	1
...	...	...	...

The following are some precautions to consider when installing a straight bore with sliding clearance coupling hub:

Insure clearance does not exceed 1 mil for shaft diameters up to 6 in.

Remove any burrs and clean components carefully before installation.

If hub sticks part way on, remove it and find the problem. Do not attempt to drive it on further with a hammer.

Install keys before the coupling is placed on the shaft.

TABLE 3.4 Guidelines for Shrink Fits on Shafts

Shaft diameter	Interference fit
1/2 to 2 in.	0.0005 to 0.0015 in.
2 to 6 in.	0.0005 to 0.002 in.
6 and up	0.0001 to 0.00035 in. per in. of shaft diameter

Straight Bore—Interference Fit with Keyways

To insure that a coupling hub can be removed once it is "shrunk" onto a shaft, proper interference fits must be adhered to. The general guideline for straight bore interference fits are found in Table 3.4

The coupling is installed by heating in an oil bath or oven to approximately 200 to 250°F and in some cases cooling the shaft simultaneously with a dry ice pack. Do not exceed 300°F or use any direct heat such as propane or oxy-acetelene torches to expand the hub, since the material properties of the shaft can be altered by direct, high-temperature concentration. Once the interference fit has been determined by measuring the shaft diameter and coupling hub bores, the temperature increase needed to expand the coupling hub to exceed the shaft diameter by 2 mils (to allow for a slide on fit) is found in equation 3.1.

$$\Delta T = \frac{i}{\alpha(d - 0.002)} \qquad (3.1)$$

where:

ΔT = rise in coupling hub temperature from ambient (°F)
i = interference fit (mils)
α = coefficient of thermal expansion (in./in.-°F) (Table 7.1)
d = coupling hub bore diameter (in.)

PULLER TOOL

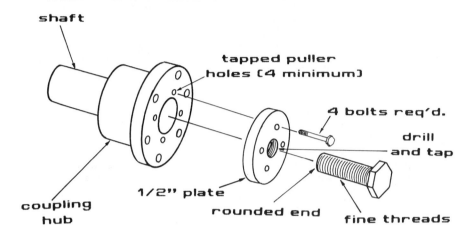

FIGURE 3.20 Coupling hub puller.

Removal of the coupling hub is accomplished by pulling the hub off the shaft with an acceptable puller mechanism and at times cooling the shaft with a dry ice pack. Shrink fit coupling hubs should always have finely threaded puller holes (preferably four) in the end of the coupling, as shown in Figure 3.20.

Bearing type pullers that push the hub off from the backside are not recommended since there is a great possibility that the puller can twist or pitch slightly, preventing a straight axial draw on the hub. For larger shaft diameters with tight interference fits, it may be necessary to apply gentle heating to the coupling hub for removal.

Splined Shaft with End Lock Nut or Locking Plate

A splined shaft and coupling arrangement is shown in Figure 3.21. There should be a slight interference fit (0.0005 in.) to prevent backlash or rocking of the hub on the shaft.

FIGURE 3.21 Splined shaft end.

Tapered Bore—Interference Fit with Keyways

Tapered shaft ends are generally used where high torques and speeds are experienced on rotating machinery necessitating a tight coupling hub to shaft fit up. The shaft end is tapered to provide an easier job of removing the coupling hub.

The degree of taper on a shaft end is usually expressed in terms of its *slope* (in./ft). The amount of interference fit is expressed in inches per inch of shaft diameter. The general rule for interference fits for this type of shaft arrangement is 1 mil per inch of shaft diameter. The distance a coupling hub must travel axially along a shaft past the point where the hub is just touching the shaft at ambient temperatures is found in equation 3.2.

$$HT = \frac{12\ I}{ST} \qquad\qquad (3.2)$$

where:

> HT = distance hub must travel axially to provide an interfer-
> ence fit equal to I
> I   = interference (mils)
> ST = shaft taper (in./ft)

Procedure for Mounting a Tapered
Coupling Hub with Keys

1. Mount bracket firmly to coupling hub and slide hub onto shaft
   end to lightly seat the hub against the shaft. Insure all surfaces are clean.
2. Measure hub travel gap HT with feeler gauges and lock nut
   down against bar as shown in Figure 3.22. Use equation 3.2
   to determine the correct axial travel needed to obtain the required interference fit onto the shaft.
3. Remove the coupling hub and puller assembly and place in an
   oven or hot oil bath to desired differential temperature. Refer to equation 3.1 to determine the required temperature rise
   to expand the coupling hub.
4. Set key in keyway and insure all contact surfaces are clean
   and burr free.
5. Carefully slide the heated coupling hub onto the shaft until
   the center measurement bolt touches the shaft end and hold
   in place until hub has cooled sufficiently.

Surface Contact

One extremely important and often overlooked consideration when
working with tapered shafts and coupling hubs is the amount of
surface contact between the shaft and hub. Due to slight machining inaccuracies, coupling hubs may not fully contact the
shaft, which results in a poor fit when the hub is shrunk or
pressed on in final assembly.

To check the surface contact, apply a thin coat of Prussian
blue paste to the inner bore of the coupling hub with your finger

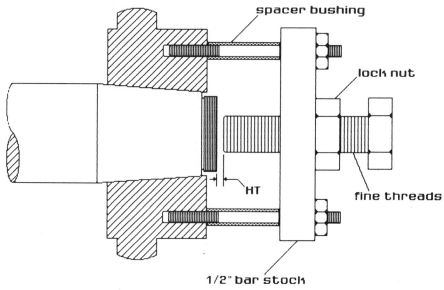

FIGURE 3.22 Measuring tool for proper interference fits on tapered shaft ends.

or a soft cloth. Slide the coupling hub axially over the tapered shaft end until contact is made, and rotate the coupling hub about 15° to transfer the paste to the shaft. Draw the coupling hub off and observe the amount of Prussian blue paste that transferred from the hub to the shaft (not how much blue came off the inside bore of the coupling hub). If there is not at least 80% contact, the fit is *not* acceptable. If the bore discrepancies are slight, it is possible to lap the surfaces with a fine grit lapping compound. Apply the compound around the entire surface contact area of the tapered shaft end lightly pushing the coupling hub up the taper and rotating the coupling hub alternately clockwise and counterclockwise through a 45° arc. Check the surface contact after 10 or 12 lapping rotations. Continue until surface contact is acceptable. However, if a "ridge" begins to develop on the shaft taper before good surface contact is made, start making preparations for remachining of the shaft and the coupling hub. It's better to bite the bullet now than try to heat the hub and put it on only to find out that it does not go on all the way, or to pick up the pieces of a split coupling hub after the unit ran for a short period of time.

Keyless Taper Bores

After working with shafts having keyways to prevent slippage of the coupling hub on the shaft, it seems very unnerving to consider attaching coupling hubs to shafts with no keys. Keyless shaft fits are quite reliable and installing hubs by hydraulic expansion methods proves to be fairly easy if you carefully adhere to installation and removal steps. Since the interference fits are usually "tighter" than those found on straight bores or tapered and keyed systems, we will review how to determine a proper interference fit.

Proper Interference Fits for Hydraulically Installed Hubs

The purpose of interference fits is twofold: 1) to prevent fretting corrosion that occurs from small amounts of movement between the shaft and the coupling hub during rotation; and 2) to prevent the hub from slipping on the shaft when the maximum amount of torque is experienced during a start up or during high running loads.

For rotating shafts, the relation among torque, horsepower, and speed can be expressed as:

$$T = \frac{63000\ P}{n} \qquad (3.3)$$

where:

P = horsepower
T = torque (in.-lbs)
n = speed (rpm)

TABLE 3.5 Allowable Torsional Stresses for Shafts

AISI no.	τ_{max} allowable torsional stress (psi)
1040	5000
4140	10000
4340	11000

The maximum amount of shearing stress in a rotating shaft occurs in the outer fibers (i.e., the fibers at the outside diameter) and is expressed as:

$$\tau_{max} = \frac{T\ r}{J} = \frac{16\ T}{\pi\ d^3} \tag{3.4}$$

where:

τ_{max} = maximum shear stress (lb/in.)
T   = torque (in.-lb)
r   = radius (in.)
d   = diameter (in.)
J   = polar moment of inertia $J = \dfrac{\pi\ r^4}{2} = \dfrac{\pi\ d^4}{32}$

The accepted "safe allowable" torsional stress for the three most commonly used types of carbon steel for shafting can be found in Table 3.5. Therefore the torsional holding requirement for applied torques is expressed as:

$$T = \frac{\tau_{max}\ \pi\ d^3}{16} \tag{3.5}$$

The amount of torque needed to cause a press fit hub to slip on its shaft is given by:

$$T = \frac{\mu\ \pi\ p\ L\ d^2}{2} \tag{3.6}$$

where:

    μ = coefficient of friction (0.12)
    p = contact pressure between hub and shaft (psi)
    L = length of coupling hub bore (in.)

The amount of contact pressure between a shaft and a coupling hub is related to the amount of interference and the outside diameters of the shaft and the coupling hub and is expressed as:

$$p = \frac{i\,E\,(DH^2 - DS^2)}{2\,(DH^2)\,(DS)} \tag{3.7}$$

where:

    p = contact pressure (lb/in.)
    i   = interference fit (mils)
    E = modulus of elasticity (30×10^6 lb/in. for carbon steel)
    DH = outside diameter of coupling hub (in.)
    DS = outside diameter of shaft (in.)

Since the shaft is tapered, dimension DS should be taken on the largest bore diameter on the coupling hub where the contact pressure will be at its minimum value, as shown in Figure 3.23.

Therefore, to find the proper interference fit between a coupling hub and a tapered shaft to prevent slippage from occurring:

1. Determine the maximum allowable torque value for the shaft diameter and the shaft material.
2. Determine the contact pressure needed to prevent slippage from occurring based on the maximum allowable torque value found in Step 1.
3. Calculate the required interference fit (solve for p in eq. 3.6 and in 3.7).

Installation of Keyless Coupling Hubs Using Hydraulic Expansion

Installing keyless taper hubs requires some special hydraulic expander and pusher arrangements to install or remove the coupling

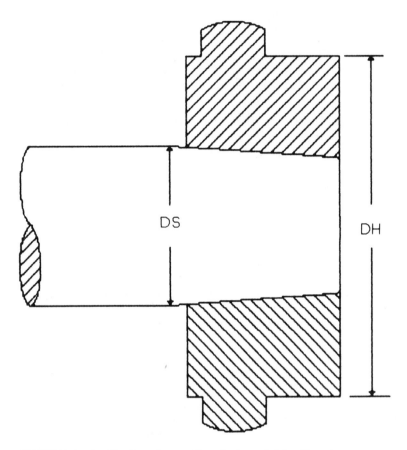

FIGURE 3.23 Shaft and coupling hub outside diameter measurement locations.

hub onto the shaft end. Figure 3.24 shows the general arrangement used to expand and push the hub onto the shaft.

Procedure for Installation of Coupling Hub Using
Hydraulic Expander and Pusher Assembly

1. Check for percentage of surface contact between coupling
    hub and shaft (must have 80% contact or better).

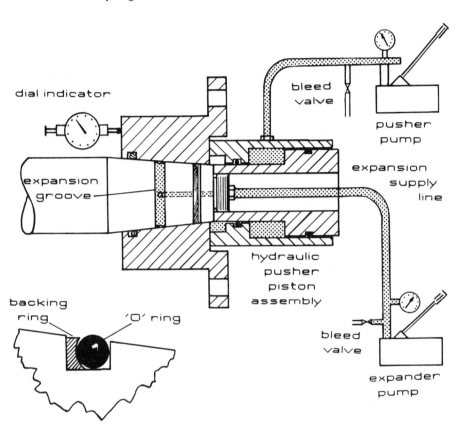

FIGURE 3.24 Hydraulic expander and pusher assembly.

2. Insure all mating surfaces are clean and that oil passageways are open and clean.
3. Install "O" rings and backup rings in coupling hub and shaft insuring that backup ring is on outside of "O" ring with respect to hydraulic oil pressure. Lightly oil the "O" rings with hydraulic fluid. Place coupling hub (and hub cover) onto shaft.
4. Install expander pump supply line to shaft end. Install the pusher piston assembly onto the end of the shaft, insuring that the piston is drawn back as far as possible. Hook up the expander pump and begin pumping hydraulic oil through

supply line to bleed any air from expansion ports and expansion groove in shaft. Once the oil has begun to seep through
the coupling hub ends, lightly push the coupling hub against
the shaft taper and begin to pump oil into the pusher piston
assembly to seat the piston against the coupling hub.

5. Place a dial indicator against the backside of the coupling
   hub and zero the indicator.
6. Begin applying pressure to the pusher piston assembly
   forcing the hub up the taper (approximately 2000 to 4000
   psig).
7. Slowly increase the pressure on the expander pump supply
   line until the coupling hub begins to move. The hydraulic
   pressure on the pusher piston assembly will begin to drop
   off as the hub begins to move. Maintain sufficient pressure
   on the pusher piston to continue to drive the hub onto the
   shaft. If the pusher piston pressure drops off considerably
   when the expansion process is underway, there is a great
   potential for the "O" rings to "blow out" the ends of the
   coupling hub, immediately seizing the hub to the shaft.
8. Continue forcing the hub up the shaft until the desired
   amount of hub travel and interference fit is attained. The
   expansion pressure will have to attain the required holding
   pressure (solve for p in eq. 3.6).
9. Once the correct hub travel has been achieved, maintain sufficient hydraulic pressure on the pusher assembly to hold
   the coupling hub in position and bleed off the pressure in
   the expansion system. Allow 15 to 20 minutes to elapse while
   bleeding to insure any trapped oil has had a chance to escape
   before lowering the pusher piston pressure.
10. Remove the pusher and expander assemblies.

Removal of the coupling hub is achieved by reversing the installation process. The key to successful installation is to take
your time and not try to push the hub up the shaft end all in
one move.

Expanding Taper Ring Shaft Connector

Another type of keyless fitup method employs two concentric, tapered thrust rings and two split inner and outer clamping rings,
as shown in Figure 3.25. As the bolts are tightened, the inner
and outer clamping rings expand radially to provide the frictional

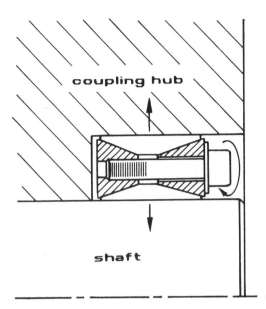

FIGURE 3.25 Expanding taper ring shaft connector. Photo courtesy of Ringfeder Corp., Westwood, N.J.

connection between the coupling hub and the shaft. Safe allowable torque capacities can be estimated by equation 3.8.

$$T = 4000 \, L \, d^2 \tag{3.8}$$

where:

> T = allowable torque capacity (in./lb)
> L = contact width of clamping rings (in.)
> d = shaft diameter (in.)

The advantage to this type of shaft and coupling hub connector is that shrinking of the coupling hubs onto the shaft by heat
or hydraulic expansion is not needed for assembly or disassembly
and no keys are needed. Precautions should be taken with peripheral velocities above 2000 fps where deformation of the coupling
hub due to centrifugal forces may cause slippage to occur. Radial
expansion of solid steel shafting can be estimated by equation 3.9.

$$\Delta e = 2 \times 10^{-11} \, U^2 \tag{3.9}$$

where:

> Δe = deformation (inches per inch of shaft diameter)
> U = peripheral velocity (inches per second)

For instance, a 4-inch diameter shaft having a 2000 fps peripheral velocity will expand approximately 3 mils. If the locking ring assembly is 5 inches in outside diameter as it clamps
against the coupling hub, an additional 0.8 mils of expansion
will occur at the split ring contact surfaces.

Threaded Shaft End Connections

Although not extensively used in industry, threaded shaft end
connections have successfully been applied primarily on refrig-
eration compressors with shaft diameters ranging between 1 and
2 1/2 in. A flexible disc coupling (see Fig. 3.5) employing one
or two thin discs is threaded onto the driver shaft. The span
between the driver and the compressor is accommodated by a thin,
flexible shaft threaded at both ends, as shown in Figure 3.26.

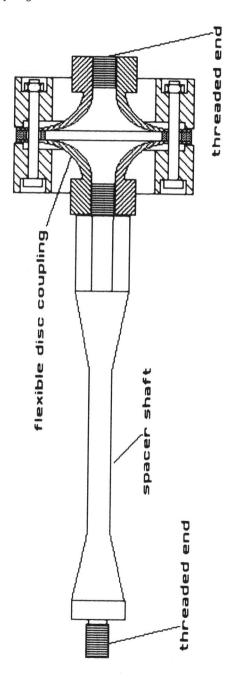

FIGURE 3.26 Flexible disc coupling with threaded shaft ends.

High torquing requirements are needed to secure the flexible coupling to the driver and spacer shafts. Expansion of the female threads can occur at high running speeds (see eq. 3.9).

BIBLIOGRAPHY

AGMA Standard, "Nomenclature for Flexible Couplings." Amer. Gear Mfgs. Assoc. AGMA 510.02 (1330 Massachusetts Ave. N.W., Wash. D.C., 20005) Aug. 1969.

"Alignment Loading of Gear Type Couplings." Bently Nevada Applications Notes no. (009)L0048. Bently Nevada Corp., Minden, Nevada (March, 1978).

Anderson, J. H. "Turbocompressor Drive Couplings." *Journal of Engineering for Industry* (Jan. 1962), pp. 115-123.

Bloch, Heinz P. "Why Properly Rated Gears Still Fail." *Hydrocarbon Processing* (Dec. 1974), pp. 95-97.

Bloch, Heinz P. "Less Costly Turbomachinery Uprates Through Optimized Coupling Selection." Proceedings, Fourth Turbomachinery Symposium, Gas Turbine Labs, Texas A&M University, Oct. 1975, pp. 149-152.

Bloch, Heinz P. "Use Keyless Couplings for Large Compressor Shafts." *Hydrocarbon Processing* (April 1976), pp. 181-186.

Bloch, Heinz P. "Improved Safety and Reliability of Pumps and Drivers." *Hydrocarbon Processing*, (Feb. 1977), pp. 123-125.

Calistrat, M. M. "What Causes Wear in Gear Type Couplings." *Hydrocarbon Processing* (Jan. 1975), pp. 53-57.

Calistrat, M. M. "Grease Separation Under Centrifugal Forces." A.S.M.E., paper no. 75-PTG-3, July 1975.

Calistrat, M. M. "Metal Diaphragm Coupling Performance." *Hydrocarbon Processing* (March 1977),

"Centrifugal Pumps for General Refinery Services." API Standard 610, American Petroleum Institute, Wash., D.C. (March 1971).

Counter, Louis F. and Fred K. Landon. "Axial Vibration Characteristics of Metal-Flexing Couplings." Proceedings, Fifth

Turbomachinery Symposium, Gas Turbine Labs, Texas A&M University, Oct., 1976, pp. 125-131.

DeRocker, D. E., S. Kaufman, and P. C. Renzo. "Gear Couplings." *Journal of Engineering for Industry* (Aug. 1968), pp. 467-474.

Gibbons, C. B. "Coupling Misalignment Forces." Proceedings, Fifth Turbomachinery Symposium, Gas Turbine Labs, Texas A&M University, Oct., 1976, pp. 111-116.

Gooding, Frank E. "Types and Kinds of Flexible Couplings." *Industrial Engineer* (Nov. 1923), pp. 529-533.

Goody, E. W. "Laminated Membrane Couplings for High Powers and Speeds." International Conference on Flexible Couplings, High Wycombe, Bucks, England, June 29 - July 1, 1977.

"Locking Assemblies RfN 7012 for Conveyor Pulleys." Ringfeder Corp., Bulletin no. 1 (April 1978).

Loosen, P., and John J. Prause. "Frictional Shaft - Hub Connectors - Analysis and Applications. *Design Engineering* (Jan. 1974).

Mancuso, J. R. "A New Wrinkle to Diaphragm Couplings." A.S.M.E., paper no. 77-DET-128, June 1977.

Moked, I. "Toothed Couplings - Analysis & Optimization." *Journal of Engineering for Industry* (Aug. 1968), pp. 425-434.

Oberg, Erik, Franklin D. Jones, and Holbrook L. Horton. *Machinery's Handbook*, 21st ed., Industrial Press Inc., New York.

Pahl, Gerhardt. "The Operating Characteristics of Gear-Type Couplings." Proceedings, Seventh Turbomachinery Symposium, Gas Turbine Labs, Texas A&M University, Dec., 1978, pp. 167-173.

Prause, John J. "New Clamping Device for Fastening Pulley End Discs to Shafts." *Skillings Mining Review*, *68* (June 1979).

"Ringfeder Locking Connections-Function-Design-Practical Applications." Ringfeder Corp., Bulletin no. 1/74A.

Serrell, John J. "Flexible Couplings." *Machinery* (Oct. 1922), pp. 91-93.

Shaw, G. V. "Employing Limited End Play Couplings Saves Motor Bearings - Sometimes." *Power* (Nov. 1955), pp. 120-121.

Spector, L. F. "Flexible Couplings." *Machine Design, 30* (Oct. 1958), pp. 102-128.

Wattner, K. W. "High Speed Coupling Failure Analysis." Proceedings, Fourth Turbomachinery Symposium, Gas Turbine Labs, Texas A&M University, Oct. 1975, pp. 143-148.

Wright, J. "Which Shaft Coupling is Best - Lubricated or Nonlubricated?" *Hydrocarbon Processing* (April, 1975), pp. 191-193.

Zurn, Frank W. "Crowned Tooth Gear Type Couplings." *Iron and Steel Engineer* (Aug. 1957), pp. 98-116.

4
Alignment Preparation

Successfully aligning rotating machinery is directly proportional to the amount of effort put forth in the preparation and planning phases of the job. The entire philosophy of this book is to provide the means for reducing the amount of time it takes to perform an alignment job on rotating machinery and to increase the accuracy of the alignment between the shafts for improved equipment operation. There are no short cuts or simple solutions, only good planning in having the right ingredients at the right time in the right place. A good rule of thumb is, expect and look for the unexpected.

Most important, don't just prepare yourself, prepare the people you will be working with on the job. There are no single heroes, just a group of heroic individual contributors working together to achieve a common goal.

The successful alignment procedure is one that balances three key elements: equipment, tools, and people to produce the philosophy mentioned above. Unless you do alignment work

FIGURE 4.1 Modified "X-mas tree" bracket.

on a weekly basis, it is a good idea to have a preparation checklist similar to the one shown in the appendix to this book. The
checklist reviews each key item, identifies the necessary steps,
and accounts for as many pitfalls as could occur during a typical alignment task. Revise this list each time you learn something new and record those unexpected pitfalls to prepare yourself better next time.

TYPES OF ALIGNMENT BRACKETS

There is no one type of alignment bracket that works for all types
of alignment conditions. In most cases, it is best to fabricate exactly what you need in your machine shop. Figures 4.1, 4.2, and
4.3 show some examples of brackets that can be used.

FIGURE 4.2 Long span "X-mas tree" bracket.

MEASURING BRACKET SAG

As the bracket is rotated from the 12 o'clock to the 6 o'clock position, gravitational force will have a tendency to distort the bracket slightly from the weight of the indicator and the frame span bar itself. This sag is one of the causes of erroneous readings but can be compensated for if the sag characteristics of the bracket and indicator are known. Mount the indicator-bracket arrangement to a 4 or 6-inch diameter pipe supported on roller stands, with the indicator spanning the same distance that it would span on the machinery being aligned. If the bracket mounts to the face of a coupling hub rather than clamping around the circumference of the coupling hub or shaft (as shown in Fig. 4.4), weld a suitable rigid post on the pipe to simulate the bracket attachment.

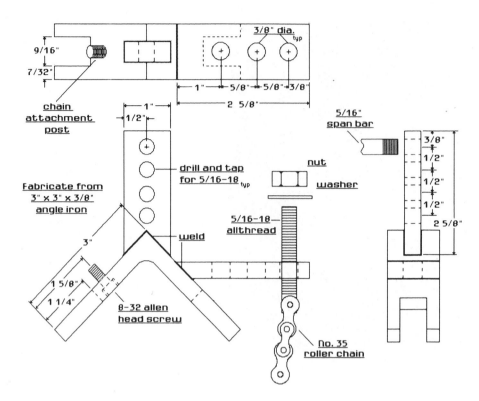

FIGURE 4.3 General purpose alignment bracket.

The sag readings will always have a negative value. When compensating for this, be sure to add the sag value to the field readings (refer to Fig. 5.9 for an example of this).

As a rule of thumb, 30 mils of sag should be the maximum allowable amount in a bracket assembly. Counterweights can be used to help compensate for this, as shown in Figure 4.4.

MEASURING SHAFT OR COUPLING HUB RUNOUT

This check is often overlooked during the alignment process and can cause considerable frustration when attempts are made to

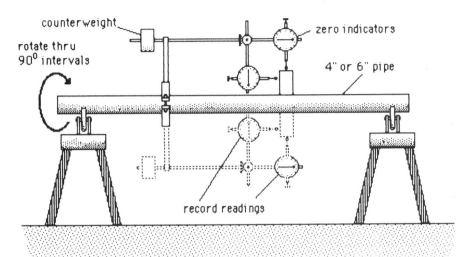

FIGURE 4.4 Bracket sag measurement setup.

align the shafts. The rule of thumb is not to exceed 2 mils of runout anywhere along the length of the shaft or one-half mil on the face of the coupling hub. If these values are exceeded, determine whether the problem is in the shaft or in the coupling hub and correct it. It is possible to compensate for this runout during the alignment process if both shafts are rotated together and the alignment bracket is put in the same position on the shaft (should it have to be removed for reverse indicator readings). If the runout is ignored by substantially greater than 2 mils, the rotational axis of the shafts may be coincident, but the excessive runout will usually cause a high once per rev vibration signal appearing as if the coupling is out of balance. Adding balance correction weights to compensate for this problem is not the correct solution (see Fig. 4.5).

PROPERLY ROTATING A SHAFT

After the alignment bracket has been fabricated and mounted on one of the shafts to begin taking alignment readings, the shaft

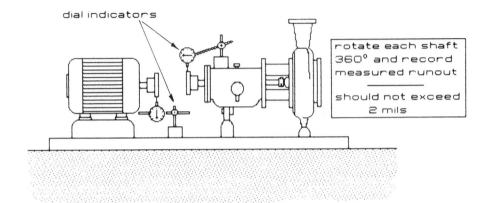

FIGURE 4.5 Measuring runout on coupling hubs.

and bracket arrangement must be rotated through a 360° arc to
record its relative position with respect to the other shaft. It is
very tempting to grab hold of the bracket frame and begin turning. This can cause erroneous readings when the bracket frame
distorts as the shaft is being rotated. The proper way to rotate a
shaft is to use one of the tools shown in Figures 4.6 through 4.8.

The Parmalee wrench in Figure 4.6 can be purchased in three
different sizes, and each size can accept a variety of girth link-
ages to grip outside diameters from 3/4 to 9 in. in 1/8 in. increments.

The chain wrench and nylon strap wrenches in Figures 4.7
and 4.8 may occasionally slip on a highly polished surface commonly found on shafts. To prevent this, wrap a 5 mil thick piece
of brass shim stock or some 600 grit emery cloth around the shaft
before attaching the wrench. *Note:* When rotating any piece of
turbomachinery, always be sure there is lubricant on the bearings to prevent damaging the bearing and shaft.

MOUNTING A DIAL INDICATOR CORRECTLY

Since the dial indicator is the primary means of measuring the
position of two shafts when monitoring the movement of machine

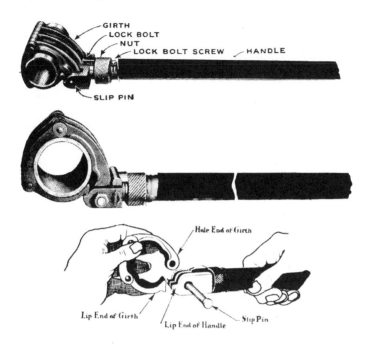

FIGURE 4.6 Parmalee wrench. Courtesy of Parmalee Wrench Co.,
Inc., Harrison, N.J.

FIGURE 4.7 Chain wrench. Courtesy of Reed Mfg. Co., Erie,
Pa.

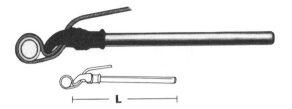

FIGURE 4.8 Nylon strap wrench. Courtesy of Reed Mfg. Co.,
Erie, Pa.

casings, it is imperative that it be mounted properly to obtain the
most accurate readings possible.

When taking the shaft-to-shaft alignment readings as explained
in Chapter 5, the indicator stem must be perpendicular to the surface where the reading is being taken. Figure 4.9 illustrates a

FIGURE 4.9 Improperly mounted dial indicator.

FIGURE 4.10 Properly mounted dial indicator.

dial indicator stem improperly mounted with it pitched slightly
and resting against the coupling hub cover. In this particular
case, both shafts will be rotating together through the four reading points at 3, 6, 9, and 12 o'clock on the coupling hub. However, if the hub cover and the driving mechanism were removed,
allowing the indicator stem to track along the circumference of
the coupling hub, the indicator stem would probably drop into
the allen head set screw, rendering the readings useless after
the indicator twists as it passes by the obstruction.

Figure 4.10 shows the correct position of a dial indicator with
its stem perpendicular to the reading surface and away from any
obstructions. The only potential hazard in this arrangement is
using a magnetic base stand instead of a rigidly attached bracket,
such as that shown in Figure 4.3.

FIGURE 4.11 Indicator stem not touching coupling hub at bottom.

Another potential problem can occur if the indicator stem does not touch the reading surface as it traverses to the bottom of the coupling hub. Figure 4.11 illustrates this point.

When monitoring lateral movement of machine casings, the indicator must be positioned to record the desired amount of movement

FIGURE 4.12 Indicator stem pitched at an angle and out of line with foot bolt.

accurately. Figure 4.12 shows the indicator stem pitched at an angle and not in line with the foot bolt. Figure 4.13 shows the correct mounting position.

FIGURE 4.13 Indicator stem correctly mounted.

BIBLIOGRAPHY

Dodd, V. R. *Total Alignment*. Petroleum Publishing Co., Tulsa, Ok., 1975.

Dreymala, James. *Factors Affecting and Procedures of Shaft Alignment*. Technical and Vocational Dept., Lee College, Baytown, Texas, 1970.

Essinger, Jack N. "A Closer Look at Turbomachinery Alignment." *Hydrocarbon Processing* (Sept. 1973).

Mannasmith, James, and John D. Piotrowski. "Machinery Alignment Methods and Applications." Vibration Institute Meeting, Cincinnati Chapter, Sept. 1983.

Murray, Malcolm G. "All the Facts About Machinery Alignment." *Hydrocarbon Processing* (Oct. 1974), pp.

Piotrowski, John D. "Alignment Techniques." Proceedings, Machinery Vibration Monitoring and Analysis Meeting, New Orleans, La., June, 1984.

5

Methods for Measuring Misalignment in Rotating Machinery

After completion of all the necessary equipment and foundation system checks, such as soft foot, clean feet, good shim packs, etc., the next step is to begin measuring the position of one shaft with respect to the other. It is amazing how accurately we can see even slight misalignment when using simple tools like a straightedge. But as the shafts begin to come more closely into line, it is easy to become overconfident with our eyesight and stop the procedure right there. I have seen many mechanics consider two pieces of equipment well aligned after laying a 6-inch pocket rule across the top and sides of the coupling hubs and if no gap is seen, bolt the coupling in, put the safety shroud on, and walk away.

The rough alignment steps will only get the equipment close enough to allow the dial indicators to stay within their allowable stem travel. To achieve the required accuracy for aligning shafts, dial indicators *must* be used during the alignment process. This chapter will explain three of the more popular methods of using dial indicators and some of the pitfalls you should be aware of when using these various techniques.

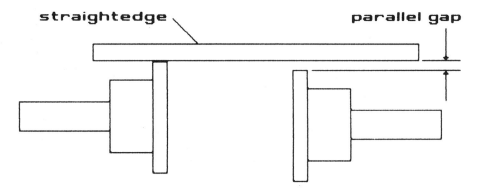

FIGURE 5.1 Straightedge for "rough" alignment.

"ROUGH" ALIGNMENT METHODS

The straightedge is quite useful in determining parallel gaps, as
shown in Figure 5.1. This only works, however, if the coupling
hubs are of the same diameter. If they are not, and the shafts
are the same diameter, use the straightedge directly on the shafts
before the hubs are installed onto the shaft ends.

Inside micrometers can be used (Fig. 5.2) to determine the
pitch or angularity of the shafts. This is probably a good time

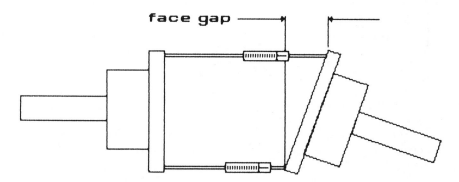

FIGURE 5.2 Using inside micrometers for "face" gaps.

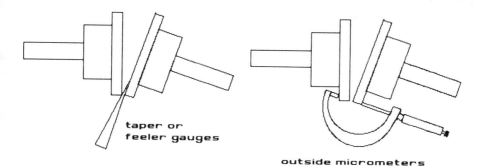

taper or
feeler gauges

outside micrometers

FIGURE 5.3 Measuring closely spaced shafts.

to make a preliminary axial spacing check. Refer to Chapter 8
for more information on axial spacing between shafts.

If the shafts are close together, taper or feeler gauges can
be used to determine the angularity and the spacing between
the shafts, as shown in Figure 5.3.

THE FACE-PERIPHERAL DIAL INDICATOR METHOD

Probably the most widely used and oldest dial indicator method
is the face-peripheral (or face and rim) dial indicator method
shown in Figure 5.4. It is a good idea when using this method
to take the face readings on the largest possible diameter on
the face of the coupling hub and be sure to measure and record
this distance for calculating the equipment moves to improve the
shaft alignment (see Chapter 6).

ADVANTAGES

Only one shaft has to be rotated
Good for large diameter coupling hubs where the shafts are close
    together
Easier to visualize the shaft positions

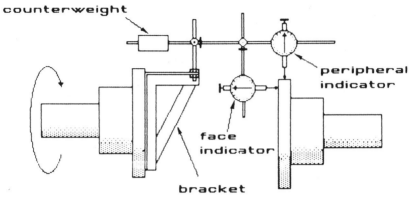

PROCEDURE

1 Attach bracket firmly to one shaft
   and position indicators on face and
   periphery of other shaft.
2 Zero the indicators at the
   12 o'clock position.
3 Slowly rotate the shaft and bracket
   arrangement through 90 degree
   intervals stopping at the 3, 6, & 9
   o'clock positions. Record each
   reading (plus or minus).
4 Return to the 12 o'clock position
   to see if the indicators re-zero.
5 Repeat steps 2 thru 4 to verify
   the first set of readings.

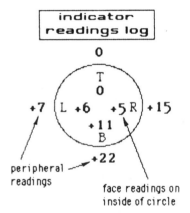

FIGURE 5.4 Face-peripheral method.

DISADVANTAGES

Difficult to obtain accurate face readings if rotor has hydrody-
      namic thrust bearings due to axial float
Most often requires removal of coupling spool
More complex to graph for alignment movement calculations

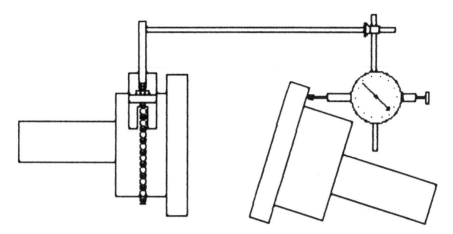

FIGURE 5.5   Taking face readings on the backside of the coupling hub.

If the shafts are too close together to be able to get a dial indicator to read the inside face of the coupling hub, readings can
be taken on the backside of the hub, as shown in Figure 5.5. When
recording the readings, change the sign (positive to negative or
vice versa) due to the reverse pitch of the reading surface.

REVERSE INDICATOR METHOD

The reverse indicator method is slowly becoming the more popular method for taking shaft alignment readings and is illustrated
in Figure 5.6.

ADVANTAGES

Geometrically more accurate than face peripheral method (readings are obtained on larger "measurement triangles")
The accuracy of the readings is not hampered by axial float of
   the rotors when turning shafts to obtain the indicator readings

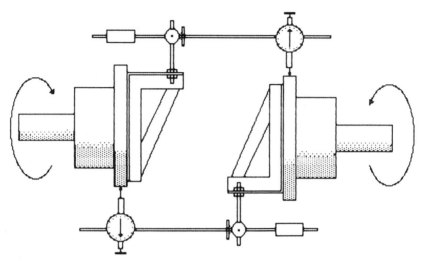

PROCEDURE

1 Attach bracket(s) to shaft(s) and
   position indicator(s) on periphery
   of other shaft(s).
2 Zero the indicator(s) at the
   12 o'clock position.
3 Slowly rotate the shaft and bracket
   arrangement(s) through 90 degree
   intervals stopping at 3, 6, & 9 o'clock.
   Record each reading (+ or -).
4 Return to 12 o'clock position to see if
   indicator(s) re-zero.
5 Repeat steps 2 through 4 to verify
   each reading.
6 If one bracket was used, mount bracket
   on other shaft and repeat steps 1 through 5.

OR to ON

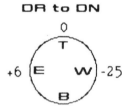

ON to OR

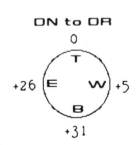

FIGURE 5.6 Reverse indicator method.

Easier to plot dynamic movement of machinery on a graph
Possible to keep the coupling spool in place when taking the
    readings

DISADVANTAGES

Both shafts have to be rotated
Should not be used on close-coupled shafts where the distance
    between the shafts is less than the coupling hub diameter
Difficult to obtain readings on extremely long shaft-to-shaft
    distances (e.g., cooling tower fan-drive systems)

SHAFT-TO-COUPLING SPOOL METHOD

In some cases it is extremely difficult, if not impossible, to fabricate an alignment bracket capable of spanning long distances
between two shafts, such as found on cooling tower fan-drive
systems. Accurate alignment can still be accomplished by using
dial indicators positioned to read a point on the coupling spool,
as shown in Figure 5.7.

TIPS FOR GETTING GOOD
ALIGNMENT READINGS

When performing any of the dial indicator methods just reviewed,
there is always a possibility that the readings obtained do not
accurately reflect the true position of one shaft with respect to
the other. In real life, a variety of problems emerge that need
to be considered when taking the alignment readings. In the
ideal situation, the readings should adhere to the *validity rule*
that states that the sum of the two side readings should be equal
to the bottom reading, as shown in Figure 5.8.

However, the ideal situation will rarely happen and the causes
for the readings not to follow the validity rule can be one or more
of the following reasons:

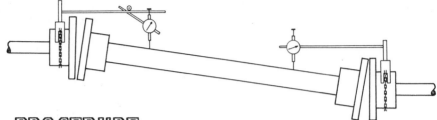

PROCEDURE

1 Attach bracket(s) to shaft(s) and position indicator at some point along length of coupling spool with indicator stem touching outside diameter of spool.
2 Zero the indicator(s) at the 12 o'clock position.
3 Slowly rotate the shafts and coupling spool assembly through 90 degree intervals stopping to record the dial indicator readings at the 3, 6, & 9 o'clock positions (+ or -).
4 Return to the 12 o'clock position to check if indicator returns to zero.
5 Repeat steps 2 through 4 to verify the readings.
6 If one bracket was used, mount the bracket on the other shaft and repeat steps 1 through 5.

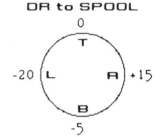

OR to SPOOL

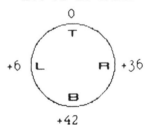

ON to SPOOL

FIGURE 5.7 Shaft-to-coupling spool method.

Sources of Face Reading Inaccuracies

Axial Float of the Shaft Being Rotated

As the shaft is being rotated, there is a tendency for the shaft, bracket, and indicator assembly to move toward or away from the other shaft. This is particularly troublesome with rotors that have hydrodynamic-type (Kingsbury) thrust bearings. One

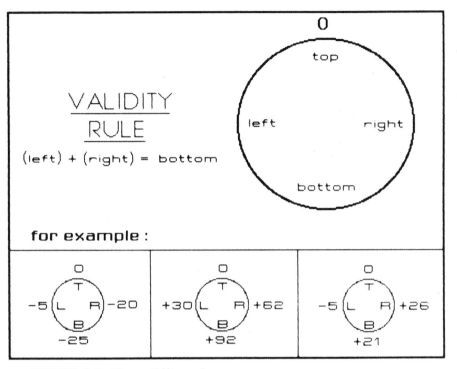

FIGURE 5.8 The validity rule.

way to eliminate this problem is to force the shafts to seat against the thrust bearings at each 90° interval. For light rotors, this can be accomplished by hand and with heavier rotors, by lightly forcing the two shafts apart with a hydraulic jack. It is a good idea to mark the shaft with a piece of tape or a felt tip pen, near the seal area, for instance, just to keep a reference when turning through the 90° angles.

Bracket Sag

Refer to Chapter 4 on how to measure this sag. Once the sag has been determined, compare your field readings with the sag readings to determine the true face readings, as shown in Figure 5.9.

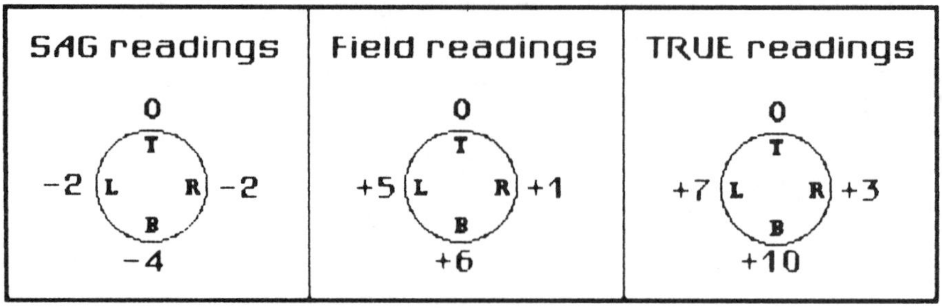

SAG readings	Field readings	TRUE readings
0 T −2 (L R) −2 B −4	0 T +5 (L R) +1 B +6	0 T +7 (L R) +3 B +10

FIGURE 5.9 Adjusting the readings to compensate for bracket sag.

"Sticky" Indicator Stem

Even the best of indicators stick and the easiest way to compensate for this is to tap the bracket or indicator lightly with your
hand to eliminate the stem gear from binding. If this problem
is severe, replace the indicator.

Not Taking "True" 90° Readings

Erroneous readings will occur if the indicator stem is slightly off
from the exact 3, 6, 9, and 12 o'clock positions. Put reference
marks on the shaft with a felt tip pen near where the dial indicator stem will be touching the coupling hub. Use a plumb bob
for the 12 and 6 o'clock points and a right-angled triangle for
the 3 and 9 o'clock points (also see Fig. 6.6).

Loose Bracket Assembly

Insure that the bracket and dial indicator are rigidly mounted.

Bent Shaft or "Skew" Coupling Hub Bore

Refer to Figure 4.5 for measuring runout on coupling hubs and
shafts.

Indicator Stem Does Not Touch Reading
Surface of Coupling Hub

At some point in the rotational process, the indicator stem may not be touching the reading surface requiring the indicator to be repositioned. Watch the indicator through the entire 360° of rotation to be sure the stem touches at all times and that your reading was 130 mils (not 30 mils) should the indicator needle swing completely around once.

Sources of Peripheral (Circumference) Reading Inaccuracies

Bracket Sag

Handle the same way as the face readings.

Sticking Indicator Stem, Loose Bracket, Not
Taking "True" 90° Readings

Handle the same way as mentioned in the face reading tips.

Tracking an "Elliptical" Reading Surface

If the shafts have a severe amount of angular misalignment, the indicator stem is tracking an elliptical path (rather than a circular one), as shown in Figure 5.10. There is no way to correct this, and in many cases the shafts must have an angular relationship due to compensation for certain types of thermal growth of the equipment. Thermal growth will be explained in Chapter 7, but for now it is important to recognize that this phenomenon exists if only to explain the inaccuracies of the readings.

When taking dial indicator readings in the field, it is important to double-check the indicator readings by rotating twice around the coupling hub to verify the initial set of numbers. If the readings do not repeat with ±10% of the original value, take another set of readings. If the readings do repeat within ±10%, average the readings, as shown in the example in Figure 5.11.

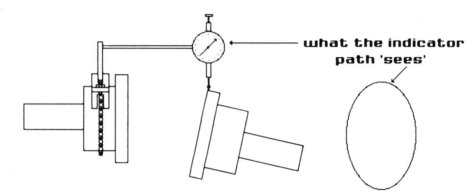

FIGURE 5.10 Elliptical tracking path seen by indicator.

THE NEXT GENERATION OF ALIGNMENT
MEASURING SYSTEMS

From relying on the accuracy of our eyesight, to using straight-
edges and feeler gauges, and then eventually to shaft brackets
and dial indicators, the art of measuring machinery shaft positions has been continually refined to improve accuracy and reduce the amount of time required to achieve acceptable machinery
alignment. It was inevitable, however, that we would not stop
there, particularly in light of the technological explosion in elec-
tronics. Now with the advent of the microprocessor chip and
the laser, it is possible to decrease further the amount of time
it takes to obtain the desired shaft alignment accuracy. These
two recently developed systems have taken the first steps toward a new generation of alignment measuring systems.

Optalign—The Laser-Optic
Measurement System

The laser system consists of an adjustable prism unit which attaches to one of the machinery shafts, a laser/detector mechanism which attaches to the other machine shaft, and a operator

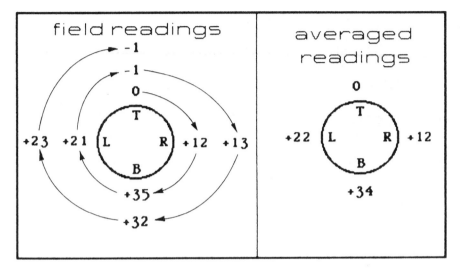

FIGURE 5.11 Rotating twice around the shaft to check the readings and average the results.

entry keypad/liquid crystal display that accepts inputs from the
laser/detector by a connecting data cable as shown in Figure
5.12.

The operating principle of the Optalign system is similar to
the face-peripheral method for measuring shaft alignment. After
the laser/detector and prism are mounted onto the two shafts,
the prism can be adjusted both laterally and vertically to center
the reflected laser beam back onto the detector. The face (i.e.,
angular) and circumferential readings are obtained by generating
a coherent beam from an infrared-emitting semiconductor laser
and reflected back into the detector via a prism, as shown in
Figure 5.13.

After the laser/detector and prism are mounted onto the
shafts, the laser beam is zeroed on the detector by mechanically
positioning the prism laterally and vertically. A hand-held laser
detector device is provided to help locate the beam if it is exces-
sively outside the range of the detector target. As the prism
is moved, the liquid crystal screen shows the amount that the

FIGURE 5.12 Laser-optic shaft alignment system. Photo courtesy of Prüftechnik Dieter Busch and Partner GmbH & Co., Ludeca, Inc., Miami, Fl.

reflected laser beam is off center on the photosensitive detector target by displaying numerical values of the beams position. Once centered, the entire drive train is rotated together (with the coupling intact) and readings are taken at the 12, 3, 6, and 9 o'clock positions. The operator then enters distances between

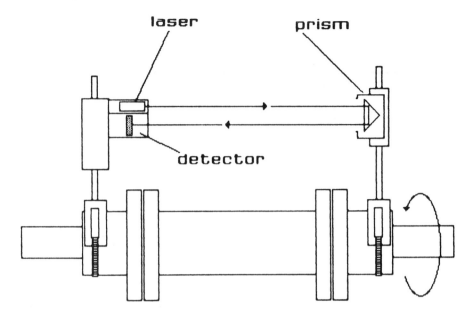

FIGURE 5.13 Operating principle of the laser system.

the laser and prism, the laser and the inboard foot of the unit chosen to be moved, and the distance between the inboard foot and the outboard foot of the machine. The amount and direction of movement of both the inboard and outboard feet are visually displayed on the screen for both vertical and horizontal adjustments. If the laser beam falls outside the range of the detector, the shafts can be rotated backwards until the displayed readings fall within the linear range of the detector. An interrupt function can be entered at this point and the prism can be "rezeroed" without having to go back to the 12 o'clock position and start the process over.

ADVANTAGES

Quick set-up time and can be mounted on a wide variety of shaft arrangements.

The coupling is kept in place when obtaining the readings.
Spans up to 6 feet between the laser, and prism can be measured.
Extreme accuracy possible under ideal environmental conditions.
Axial "float" of the shafts do not affect the reading accuracy.
User friendly interface with the operator keypad.
Visual representation of the misalignment condition with automat-
     ically calculated movement values.

DISADVANTAGES

Torsionally "loose" couplings, such as gear, chain, and metallic
     grid type designs, must be locked to prevent backlash dur-
     ing rotation.
Accuracy reduced in the presence of excessive heat or steam
     sources near the laser and/or prism.
Calculator/display system is set up to determine required moves
     for only one of the two machine elements.
Cost of system and delicacy of each component.
Periodic factory recalibration of system required to maintain ac-
     curacy.
Cannot overlay a projected line for a minimum movement calcula-
     tion (refer to Chapters 6 and 7).

The Lineax Shaft-to-Shaft Alignment Instrument

As shown in Figure 5.14, the Lineax system consists of: two tri-
dimensional sensing units that are rigidly attached to each shaft,
a rigid signal extension rod that attaches one sensing unit to
the other, and an operator interface display system connected
to both sensing units. The sensing units can be placed on different diameter shafts, and the system comes equipped with various lengths of signal extension rods from 8 to 32 in. (in 2-in. increments).
    Once the sensing units have been attached to each shaft, the
operator places the sensing units in the upright position using
level indicators on each sensor. The stainless steel signal extension rod is then placed between the two sensing units and
signal feedback cables are attached between each sensor and
the operator interface display and readout device. Dimensions
between the sensors and the inboard and outboard feet are entered into the readout/display unit, and without having to rotate

FIGURE 5.14 Lineax alignment system. Photo courtesy of Spring-Mourne, Inc., Lineax Instruments Division, Brookfield, Wis.

the drive train, the required vertical and horizontal moves are calculated and displayed on the display system.

ADVANTAGES

Quick set-up time.
Shafts do not have to be rotated to obtain readings.
The coupling can be rigidly bolted together or left uncoupled
     when obtaining the readings.
Extreme accuracy possible under ideal environmental conditions.
User friendly interface with the operator keypad.
Visual representation of the misalignment condition with automat-
     ically calculated movement values for either the driver or
     driven unit.
Automatically adjusts for different shaft diameters (within the
     range of the specific model).

Allows for monitoring of the shaft positions during the alignment
moves.
Can be used to find and correct a soft foot condition quickly.

DISADVANTAGES

One model cannot accommodate a wide variety of different shaft
diameters.
Accurate measurement of shaft diameters required.
Cost of system and delicacy of each component.
Cannot overlay a projected line for a minimum movement calcula-
tion (refer to Chapters 6 and 7).

BIBLIOGRAPHY

Dodd, V. R. *Total Alignment*. Petroleum Publishing Co., Tulsa,
Ok., 1975.

Dreymala, James. *Factors Affecting and Procedures of Shaft
Alignment*. Technical and Vocational Dept., Lee College,
Baytown, Texas, 1970.

Durkin, Tom. "Aligning Shafts, Part I - Measuring Misalignment."
Plant Engineering (Jan. 1979), pp. 86-90.

Murray, M. G. "Choosing an Alignment Measurement Setup."
Murray and Garig Tool Works, Baytown, Texas, personal
correspondence, Oct. 12, 1979.

Murray, M. G. "Optalign - Laser - Optic Machinery Alignment
System - Report Following Four Month Test." Murray and
Garig Tool Works, Baytown, Texas, April, 1985.

Nelson, Carl. A. "Orderly Steps Simplify Coupling Alignment."
Plant Engineering (June 1967), pp. 176-178.

Piotrowski, John D. "The Graphical Alignment Calculator." *Machinery Vibration Monitoring and Analysis*, Vibration Institute, Clarendon Hills, Ill., 1980.

Piotrowski, John D. "Alignment Techniques." Proceedings, Machinery Vibration Monitoring and Analysis Meeting, New Orleans, La., June 1984, pp. 214-246.

"Two Step Dial Indicator Method." Bulletin No. MT-SS-04-001, Rexnord, Thomas Flexible Coupling Division, Warren, Pa., 1979.

6
Determining the Proper Machinery Moves

The reverse indicator and face-peripheral shaft measurement methods shown in Chapter 5 are fairly well known. A considerable number of people in industry today are familiar with these dial indicator methods and use them quite frequently to align rotating machinery. However, the most formidable task of the alignment process is the ability to interpret these readings and accomplish the required alignment accuracy with a minimum number of machinery moves in the shortest possible time.

As sad as it may sound, the majority of the people still use trial-and-error methods to align the equipment, and it seems to be quite difficult at times to convince many of them that there are better ways to get the job done. It wasn't until the early 1970's that a few individuals actually began to calculate the amount and direction a unit had to be moved to bring two or more equipment shafts closer in line. The graphical plotting technique began appearing out in the field, yet many of the now known "shortcuts" unique to this method were not being utilized. The mathematical equations were available to calculate the moves,

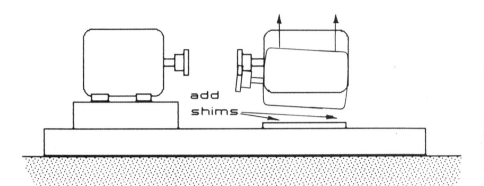

FIGURE 6.1 Vertical movement of machinery.

but figuring them out by hand took as much time as putting some
shims under the feet and seeing what effect the shim change had
on the indicator readings. With the advent of the hand-held calculator, this mathematical burden was somewhat lessened even
though there were a dozen or so equations one had to perform
to get the desired numbers crunched out. Even then, figuring
out if a negative sign meant you had to move the unit to the
west or to the east was still a problem.

With the recent introduction of the portable mini-computer
and specialized software, the required amount and direction of
required movement on one of the elements in the drive train can
be rapidly formulated and displayed to the person doing the
alignment. But the existing software limits your options to
moving just one of the drive train elements, and this is generally not the easiest and quickest solution possible. There is
also a general distrust of techniques that only show numbers
and do not provide some visual representation of the shaft positions before and after a move is made. Calculators and computers are not inexpensive tools and are usually not rugged
enough to be used in the field where they could be dropped and
damaged. When you are at a job site where the calculations have
to be performed, your hands have a tendency to pick up copious
amounts of dirt and grease that do not mix well with pushbuttons,
LED and LCD displays, and printed paper outputs.

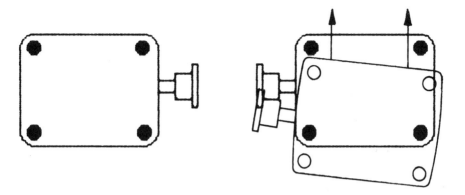

FIGURE 6.2 Horizontal movement of machinery.

By far, the graphical plotting boards shown in this chapter represent the least expensive tool for calculating the machinery moves, have the ability to display visually the position of the shafts before and after a move is made, are as accurate as any computer or calculator, and most important, allow you an unlimited number of options for moving the machinery.

Nevertheless, use the technique with which you feel most comfortable, whether it be a computer or graphical calculator or some other ingenious device you've invented.

MOVEMENT DIRECTION PLANES

The three basic planes of movement are: *vertical* (up and / or down), *horizontal* (sideways), and *axial* (shaft ends toward or away from one another) as shown in Figures 6.1 through 6.3

Chapter 8 will discuss how the axial spacing between shafts is achieved and what to watch out for on certain kinds of equipment. This chapter deals primarily with determining the movement of machinery in the vertical and horizontal planes.

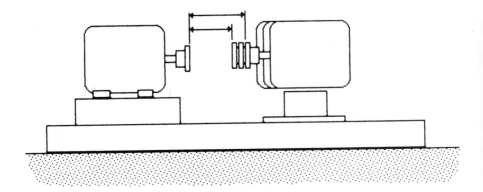

FIGURE 6.3 Axial movement of machinery.

HOW TO HANDLE THE SIDE READINGS FOR
LATERAL MOVEMENT CALCULATIONS

A lot of people have trouble with this when first learning alignment. As Chapter 5 pointed out, the dial indicators are zero at
the 12 o'clock position on the coupling hub. Figure 6.4 shows
some typical readings that may be encountered.
      The dial indicator does not necessarily have to be zeroed at
the 12 o'clock position on the coupling hub. If the dial indicator was zeroed on the right (R) side instead of the top position,
the readings obtained for the same dial indicator arrangement
would look like those in Figure 6.5.
      Just by the addition of 21 mils to each reading, Figure 6.5
was obtained. Another way to do this is to position the dial indicator at the 3 or 9 o'clock position and zero the indicator
there. Then rotate the bracket and shaft arrangement so the
indicator swings 180° around to the other side and note the
reading. An example of this can be found in Chapter 8. If
you're calculating lateral or sideways moves, the top and bottom readings are not used. Likewise, if you're calculating vertical moves (i.e., shim changes), the two side readings are not
used.

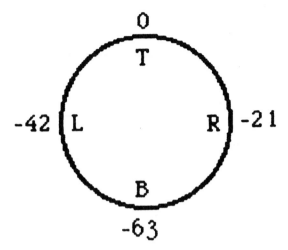

FIGURE 6.4 Example of peripheral readings taken on a coupling hub.

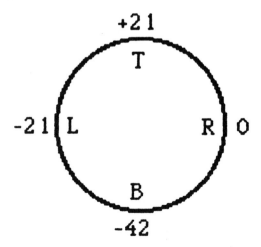

FIGURE 6.5 Zeroing the indicator on the side.

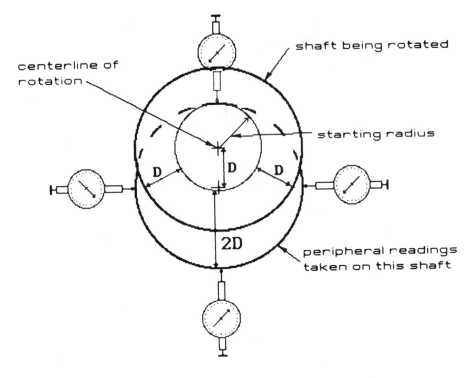

FIGURE 6.6 Axial view of the shafts where the readings are being taken on the coupling hubs.

ONE-HALF THE INDICATED READING

One of the confusing things about taking dial indicator readings on the periphery of a shaft or coupling hub is that the actual centerline offset of the shafts is only one-half the indicated reading shown by the dial indicator. To understand this phenomena better, imagine being able to look down the centerline of rotation of the shaft to which the bracket and dial indicator are attached and observe what is happening to the dial indicator stem as it travels along the periphery of the other shaft, as shown in Figure 6.6. The shaft being rotated is above the centerline of the other shaft by distance D. As the dial indicator swings from the 12 o'clock to the 6 o'clock position, the dial indicator has seen a distance of twice the actual centerline offset.

The particularly astute observer will also note another phenomenon when the indicators have been placed at the 3 and 9 o'clock positions on the reading surface. The distance from the starting radius to the point of contact on the coupling hub surface is slightly greater than distance D, also contributing to the problems mentioned in Chapter 5 on the validity rule.

BASIC MATHEMATICAL SOLUTIONS FOR CALCULATORS AND COMPUTERS

Regardless of its diameter, a shaft can be represented by a straight line coincident with its axis of rotation. Since we are dealing with two (or more) straight lines, there exists some geometric relationship between these lines that can be expressed mathematically. The principle used to generate the mathematical equations when solving for the necessary shim changes of lateral movement at the inboard and outboard feet is shown in Figure 6.7

In other words, the height of similar scalene triangles are proportional to their bases. The distance ht, for instance, would be the distance from where the indicator bracket is attached on one shaft to the point where the dial indicator stem is touching the other shaft. Distance HT would be from where the bracket is attached to the location of the inboard feet and HHT to the outboard feet. Distance bs would be the diameter on which the face readings are being taken, BS the required move at the inboard foot, and BBS the required move at the outboard foot.

It is important to recognize that the equations shown in Figures 6.8 and 6.9 will give the required movement at the inboard and outboard foot to bring the shafts directly into line and does not account for dynamic machinery movement. Chapter 7 will cover this in greater detail and will provide you with enough information to adjust the equations to "cold" offset the shaft so they move into a colinear axis of rotation when running.

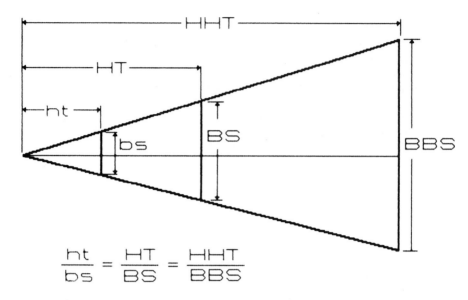

$$\frac{ht}{bs} = \frac{HT}{BS} = \frac{HHT}{BBS}$$

FIGURE 6.7 Base-to-height relationship in scalene triangles.

NOMENCLATURE

A = the distance from the outboard foot of the driver unit to where the readings are being taken on the driven unit

B = the distance from the inboard foot of the driver unit to where the readings are being taken on the driven unit

C = the distance between the dial indicator readings on each shaft. This is usually greater than the shaft-to-shaft distance

D = the distance from the inboard foot of the driven unit to where the readings are being taken on the driver unit

E = the distance from the outboard foot of the driven unit to where the readings are being taken on the driver unit

F = the reading taken on the face of the coupling hub when the dial indicator is rotated to the 6 o'clock position after being zeroed at the 12 o'clock position

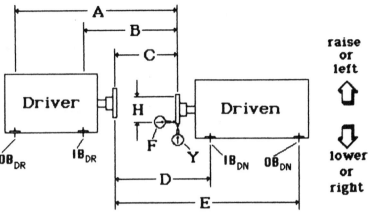

Required movement for DRIVER unit	Required movement for DRIVEN unit
$IB_{DR} = \dfrac{FB}{\sqrt{H^2 - F^2}} - (Y)$	$IB_{DN} = \dfrac{F[(D) - (C)]}{\sqrt{H^2 - F^2}} + (Y)$
$OB_{DR} = \dfrac{FA}{\sqrt{H^2 - F^2}} - (Y)$	$OB_{DN} = \dfrac{F[(E) - (C)]}{\sqrt{H^2 - F^2}} + (Y)$

FIGURE 6.8 Mathematical solutions when using the face-peripheral method. Positive (+) values for the inboard and outboard calculations indicate that the unit has to be raised (i.e., shims added) or moved to the left (i.e., lateral movements). Negative (-) values for the inboard and outboard calculations indicate that the unit has to be lowered or moved to the right.

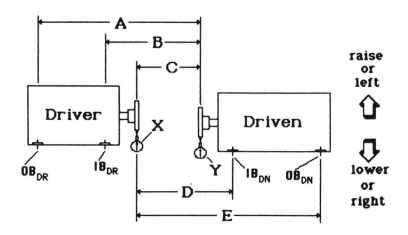

FIGURE 6.9 Mathematical solutions when using the reverse indicator method. Positive (+) values for the inboard and outboard
calculations indicate that the unit has to be raised (i.e., shims
added) or moved to the left (i.e., lateral movements). Negative
(-) values for the inboard and outboard calculations indicate that
the unit has to be lowered or moved to the right.

H = the diameter the face readings were taken on the coupling hub of the driven shaft

IB = the amount of movement needed at the inboard (i.e., coupling end) foot point that will bring the two shafts into line

OB = the amount of movement needed at the outboard foot point that will bring the two shafts into line

X = one-half the dial indicator reading taken with the alignment bracket mounted on the driven shaft and peripheral readings taken on the driver shaft

Y = one-half the dial indicator reading taken with the alignment bracket mounted on the driver shaft and peripheral readings taken on the driven shaft

GRAPHICAL ALIGNMENT

Graphical alignment is a technique that shows the relative position of two shaft centerlines on a piece of square grid graph paper by scaling down the machine dimensions from right to left and amplifying the dial indicator readings obtained from face-peripheral or reverse indicator alignment methods from top to bottom on the graph. It actually enables one to see how the shafts are misaligned in either the vertical or horizontal plane and decide whether it is easier to move one unit or the other or both to achieve the quickest alignment solution. This procedure can be used by drawing the shaft centerlines directly on to a piece of graph paper or by constructing a graphical calculator similar to the ones shown in Figures 6.10 or 6.11.

The calculators can be made to any size but the larger calculators are slightly more accurate.

The first step is to measure the following distances: foot point to foot point on both the driver and driven units, the distance from the inboard foot to where the dial indicator readings were taken on the coupling hub, and the distance between dial indicator readings. Then choose an appropriate scale to fit these dimensions along the chart centerline, as shown in Figures 6.12 and 6.13.

The calculator is used for both vertical movement (Fig. 6.14) and sideways movement (Fig. 6.15).

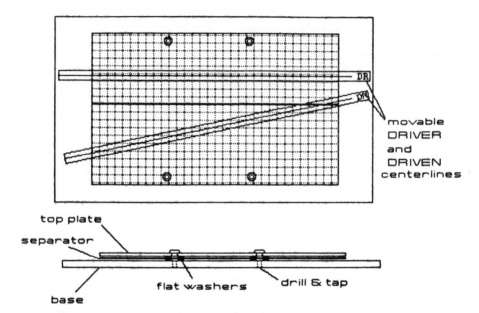

FIGURE 6.10 Graphical alignment calculator. To construct:
(1) fabricate from clear plexiglass; (2) place 10 X 10 grid graph
paper between separator and base; (3) slide movable driver
(DR) centerline between top plate and separator; and (4) slide
movable driven (DN) centerline between graph paper and separator.

GENERAL PROCEDURE USING THE REVERSE
INDICATOR ALIGNMENT METHOD

Since the graph setup is similar for both vertical or sideways
movement, the following rules will show the steps to set up the
position of both shafts as viewed from the side of the unit. This
will enable one to determine the amount of vertical movement
(i.e., shim changes) of each, or both, units to bring the cen-
ters of rotation in line.

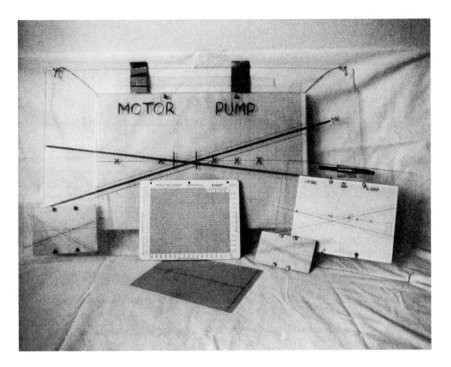

FIGURE 6.11 Various sizes and types of graphical calculators. Middle calculator courtesy of Murray and Garig Tool Works, Baytown, Tx.

Part A: Constructing the Driven
Unit Centerline on the Graph

1. If the bottom reading on the driven coupling hub was negative, start at the intersection of the graph centerline and the driven coupling hub line and mark a point directly above the graph centerline an amount equivalent to half the bottom reading using the graph division lines as 1 mil each. If the bottom reading is a large number, the divisions may be 2, 5, 10 mils or whatever will allow the complete centerline to fit on the graph. Choose

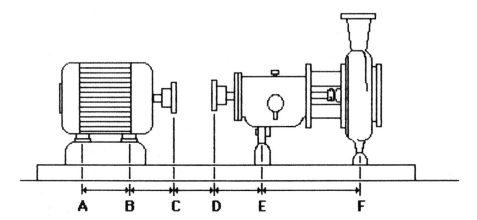

FIGURE 6.12 Take dimensions along the length of drive train.

the best scaling factor to fit the machinery shaft centerlines within the boundaries of the graph. The smaller the scaling factor
value is, the greater the accuracy of the chart. Use the same
scaling factor throughout the rest of the steps.

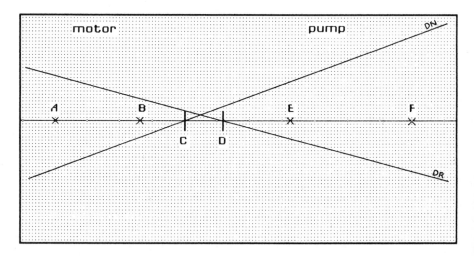

FIGURE 6.13 Scale the dimensions along the graph centerline.

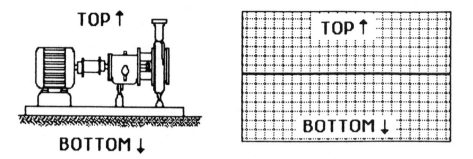

FIGURE 6.14 Using the calculator for vertical movement.

If the bottom reading was positive, start at the intersection of the graph centerline and the driven coupling hub line and mark a point directly below the chart centerline an amount equivalent to half the bottom reading.

2. Position the movable driven (DN) centerline to go through the marked point on the driven coupling hub line and the point where the graph centerline and the driver coupling hub line intersect, as shown in Figures 6.16 and 6.17.

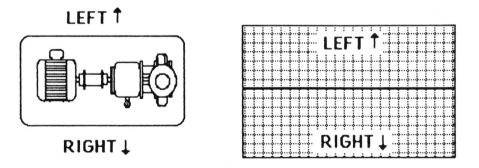

FIGURE 6.15 Using the calculator for horizontal movement.

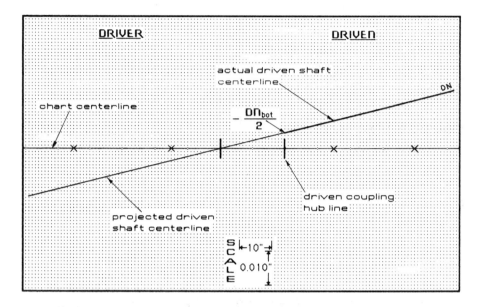

FIGURE 6.16 Position of the driven shaft when the bottom reading is negative.

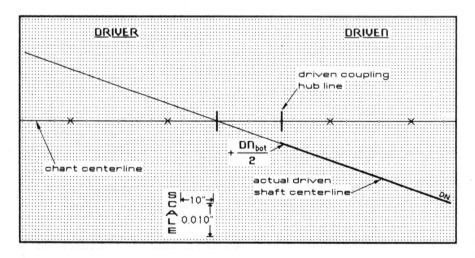

FIGURE 6.17 Position of the driven shaft when the bottom reading is positive.

Part B : Constructing the Driver
Unit Centerline on the Graph

3. If the bottom reading on the driver coupling hub was negative, start at the intersection of the graph centerline and the driver coupling hub line and mark a point directly above the graph centerline an amount equivalent to half the bottom reading.

If the bottom reading was positive, start at the intersection of the graph centerline and the driver coupling hub line and mark a point directly below the graph centerline an amount equivalent to half the bottom reading.

4. Position the movable driver (DR) centerline to go through the marked point on the driver coupling hub line and the point where the graph centerline and the driver coupling hub line intersect, as shown in Figures 6.18 and 6.19.

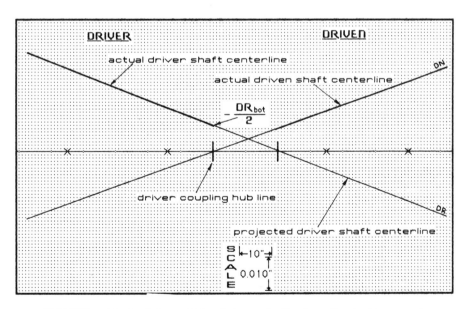

FIGURE 6.18 Position of the driver shaft when the bottom reading is negative.

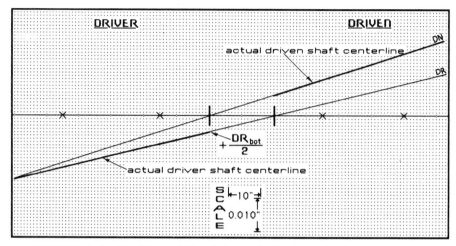

FIGURE 6.19 Position of the driver shaft when the bottom reading is positive.

Example 1. Basic Graph Calculations

Figure 6.20 shows an electric motor driving a single-stage centrifugal pump. Scale the dimensions onto the graph as illustrated in Figure 6.21.

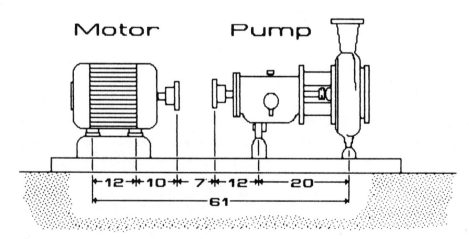

FIGURE 6.20 Motor and pump dimensions.

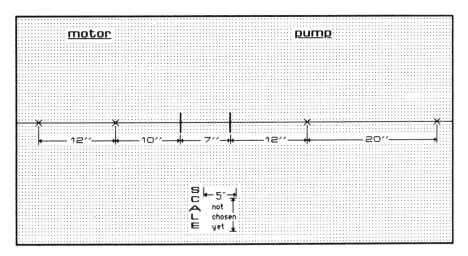

FIGURE 6.21 Drive train dimensions scaled onto the graphical chart.

Since the dial indicator readings show a considerable amount of sideways misalignment, the calculator will be set up to show the amount and direction of lateral movement needed. In most cases, it is best to improve the side-to-side readings first for two reasons: (1) it is easier to slide equipment sideways than it is to

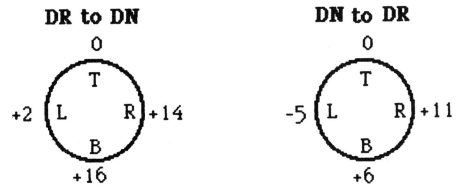

FIGURE 6.22 Reverse indicator readings taken on motor and pump.

DR to DN

O

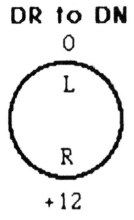

L

R

+12

DN to DR

O

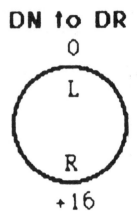

L

R

+16

FIGURE 6.23 Side readings only.

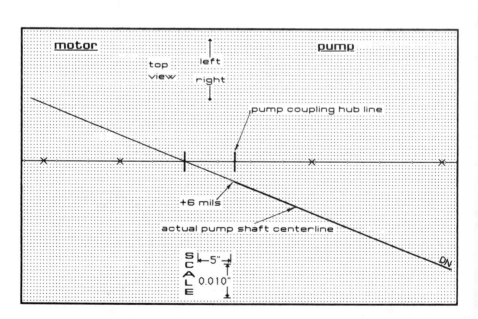

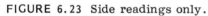

FIGURE 6.24 Constructing the pump centerline.

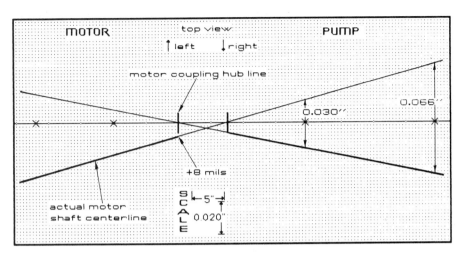

FIGURE 6.25 Constructing the motor centerline.

lift and add or remove shims; and (2) the bottom reading is more accurate when the two side readings are equal (or nearly so). For this example, the readings taken on the left side of the coupling hubs indicate that you are standing at the outboard end of the motor and looking at the pump. The top of the graph therefore corresponds to the left side of the drive train and the bottom of the graph corresponds to the right side of the two units. Zero the reading on the left side by numerically subtracting that value from the right side readings, as shown in Figure 6.23.

To construct the pump (DN) centerline, mark a point along the pump coupling hub line one-half the right side reading which is +6 mils toward the bottom of the chart (Fig. 6.24) and position the DN centerline as explained in Step 2 in the general procedure. Each small graph division from top to bottom on the graph is equal to 1 mil.

To construct the motor (DR) centerline, mark a point along the motor coupling hub line one-half the right side reading which is +8 mils toward the bottom of the chart, as shown in Figure 6.25. Position the DR centerline as explained in Step 4 in the general procedure.

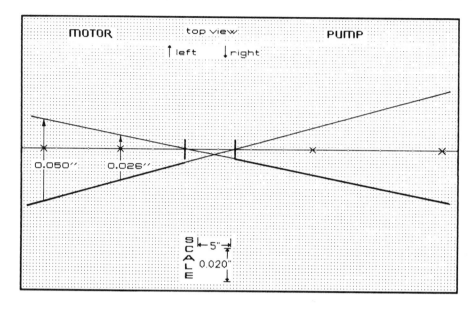

FIGURE 6.26 Moving the motor instead of the pump.

The graph now shows the relative position of each shaft as
if you were looking down on the two units from above. For this
example, let's choose to move the pump to the left to bring the
shafts into line. To do this, count the number of divisions along
the inboard foot of the pump between the pump centerline and
the projected motor shaft centerline and multiply by 0.001-in.
As shown in Figure 6.25, the pump is to be moved 0.030 to the
left at the inboard foot and 0.066 in. to the left at the outboard
foot. The same exercise could be performed on the motor if desired, as illustrated in Figure 6.26. By moving the inboard foot
of the motor 0.026 in. to the left and outboard foot of the motor
0.050 in. to the left, the same results would be achieved. These
are but two possible moves that could be made on the units to
bring them in line sideways.

The alignment in this example is not yet finished since shims
may have to be added or removed from the feet of the pump or
the motor. At this point another set of reverse indicator readings are taken and the readings yield the results shown in Figure 6.27.

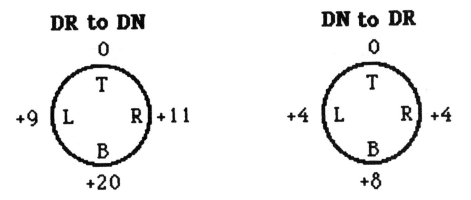

FIGURE 6.27 Reverse indicator readings after moving the equipment sideways.

The side readings are now satisfactory and the graph is now used to determine the vertical movement needed. The plotted centerlines are shown in Figure 6.28.

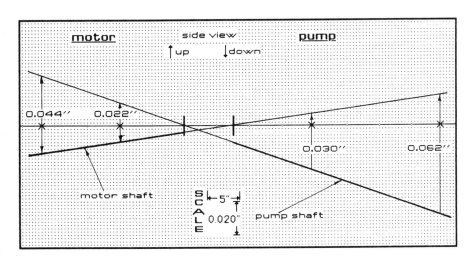

FIGURE 6.28 Position of the motor and pump centerlines for vertical movement calculations.

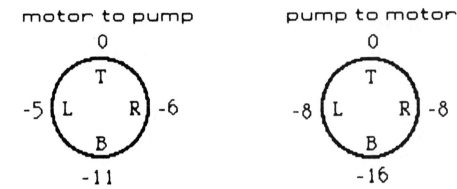

FIGURE 6.29 Reverse indicator readings.

        Decide which of the two units are to be moved. If the motor
is to be moved, 0.022-in. shims are to be added to the inboard
feet and 0.044-in. shims are to be added to the outboard feet,
to bring the two shafts directly in line. If the pump is to be
moved, 0.030-in. shims are to be added to the inboard feet and
0.062-in. shims are to be added to the outboard feet. After the
shims were added to the motor or to the pump, another set of
readings should be taken to verify the position of both shafts
to insure that the proper amount of shims were added. If the
alignment readings are satisfactory, record these readings for
future reference on the machinery data card for these two units.

Example 2. Overlaying a Projected
Line on the Graph

Using the same motor and pump arrangement as shown in Example 1, and after performing the necessary sideways movement,
the set of reverse indicator readings shown in Figure 6.29 are
taken.
        In this example, however, there are no shims under either
the motor or the pump, as both units are sitting directly on the
baseplate. The graph shown in Figure 6.30 illustrates that the
motor be lowered 26 mils at the inboard feet and 50 mils at the
outboard feet or that the pump be lowered 28 mils at the inboard

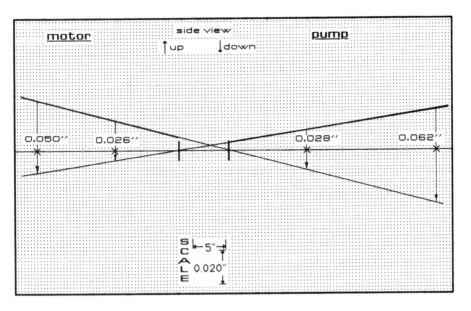

FIGURE 6.30 Position of motor and pump shaft centerlines.

feet and 62 mils at the outboard feet. The only possible way to
lower either the motor or the pump is to remove metal from the
baseplate, wasting a tremendous amount of time and labor.

Instead, project a straight line between the points of intersection of the pump and motor shafts and the outboard foot point
lines of both units. Use the two outboard feet of each unit to
"pivot" from and lift the inboard feet of each unit up, as shown
in Figure 6.31.

Using the straightedge overlay technique on the graph per-
mits an infinite number of possible moves. Projecting a straight
line between the two outboard feet, or the two inboard feet, or
one inboard foot and one outboard foot, will save a considerable
amount of time cutting shims or machining baseplates. This
method comes in handy when piping strain is a problem and
dropping or lifting a pump by 1/8 in. or so relieves the piping
stresses, as shown in Figure 6.32. It is also invaluable when
moving a unit sideways and the equipment becomes "bolt bound"
before the desired amount of movement is achieved. Another
couple of minutes spent staring at the chart and figuring out all
the possible moves proves to be a time-saver in the long run.

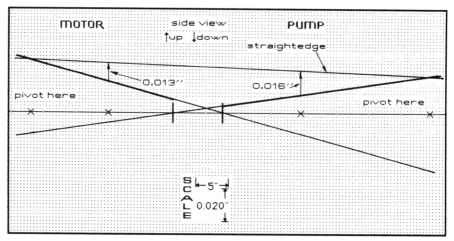

FIGURE 6.31 Using the outboard feet as "pivot" points.

Applying the Graphical Calculator when Using
the Shaft-to-Coupling Spool Indicator Method

The graphical alignment calculator can also be used to illustrate
the relative position of two shafts spaced a considerable distance
apart, such as cooling tower fan-drive systems. The only alter-
ation that is needed will be to use the coupling spool as the reference for both shafts.

Basically, just substitute Steps 2 and 4 in the general procedure (pp. 128-133) for the reverse indicator alignment method
as follows:

2. Position the movable driven (DN) centerline to go through
the marked point on the driven coupling hub line and the point
where the indicator stem touches the coupling spool and the chart
centerline intersect, as shown in Figure 6.33.

4. Position the movable driver (DR) centerline to go through
the marked point on the driver coupling hub line and the point
where the indicator stem touches the coupling spool and the chart
centerline intersect, as shown in Figure 6.34.

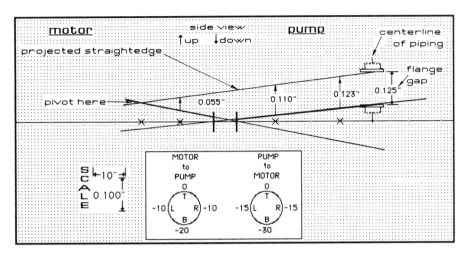

FIGURE 6.32 Adjusting the alignment to relieve piping strain on a pump.

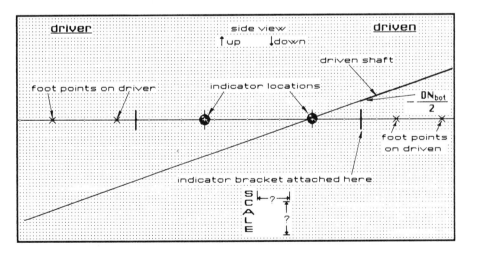

FIGURE 6.33 Constructing the driven shaft centerline.

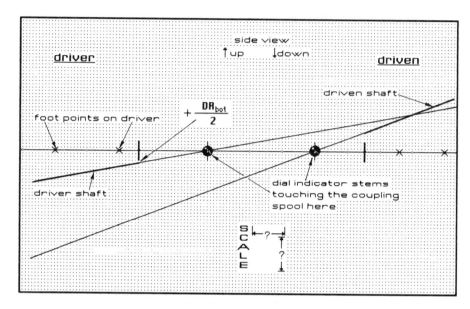

FIGURE 6.34 Constructing the driver shaft centerline.

Example 3. Using the Shaft-to-Coupling Spool
Method on the Graphical Calculator

Figure 6.35 shows an electric motor driving a right-angled gear
on a cooling tower fan-drive system. The entire motor, coupling,
and fan shafting arrangement is rotated together, and readings

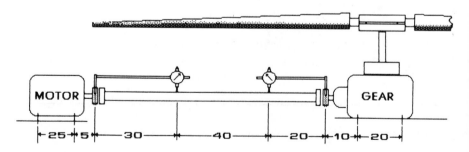

FIGURE 6.35 Dial indicator set up and dimension to scale on the
graph.

Motor to spool   Gear to spool

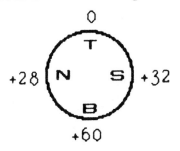
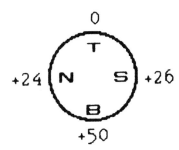

FIGURE 6.36 Initial set of dial indicator readings.

are taken on the coupling spool at 90° intervals with a bracket
and indicator attached to the motor shaft. Another set of readings is taken with a bracket and indicator attached to the fan
(gear) shaft. To determine the necessary machinery movements,
the same guidelines used for the reverse indicator procedure
will be used here for setting up the shaft positions on the graphical plot.

A set of readings is taken from the motor to the coupling
spool and from the fan gear to the coupling spool, and the
results are shown in Figure 6.36.

Based on the dial indicator readings, the motor and gear
shaft centerlines are plotted on the graph, as shown in Figure
6.37. A straightedge was projected through the inboard foot
point line of the motor and the outboard foot point line of the
gear where they respectively intersect their actual shaft centerlines. As shown, the outboard foot of the motor must be
lifted 30 mils upward and the inboard foot of the gear must
be lowered 25 mils, using one foot on each unit from which
to pivot.

The examples in this chapter have illustrated how to calculate the necessary movement of equipment to bring the shafts
directly into alignment. Movement of machinery due to temperature changes in the casings, foundation movement, or other
forces will change the position of the centerlines from "off"
to "on" line conditions. The movement of rotating machinery
and the positioning of shafts during the alignment process
to compensate for this movement will be covered in the next
chapter.

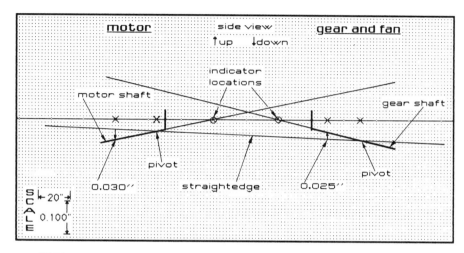

FIGURE 6.37 Position of motor and fan drive shafts showing one
possible alignment solution.

BIBLIOGRAPHY

Dodd, V. R. *Total Alignment*. Petroleum Publishing Co., Tulsa,
    Ok., 1975.

Dreymala, James. *Factors Affecting and Procedures of Shaft
    Alignment*. Technical and Vocational Dept., Lee College,
    Baytown, Texas, 1970.

Durkin, Tom. "Aligning Shafts, Part II - Correcting Misalign-
    ment." *Plant Engineering* (Feb. 1979), pp. 102-105.

Jackson, Charles. *The Practical Vibration Primer*. Gulf Pub-
    lishing Co., Houston, Texas, 1979.

Machinery Alignment Plotting Board. Murray and Garig Tool
    Works, Baytown, Texas, 1978.

Malak, Steven P., and Donald W. Solomon, "An Analysis of Sta-
    tionary 3-Dimensional Shaft Alignment Instrumentation Com-
    pared with Rotated 2-Dimensional Shaft Alignment Measure-
    ments Obtained Currently by Dial Indicator Techniques,"
    SMI-LI-TR1, Lineax Instruments Div., Spring-Mornne, Inc.,
    Brookfield, Wisc., August, 1985.

Mannasmith, James, and John D. Piotrowski. "Machinery Alignment Methods and Applications." Vibration Institute meeting, Cincinnati Chapter, Sept. 1983.

Murray, M. G. "All the Facts About Machinery Alignment." *Hydrocarbon Processing* (Oct. 1974).

Murray, M. G. "Out of Room? Use Minimum Movement Machinery Alignment." *Hydrocarbon Processing* (Jan. 1979).

Piotrowski, John D. "The Graphical Alignment Calculator." *Machinery Vibration Monitoring and Analysis*. Vibration Institute, Clarendon Hills, Ill., 1980.

Piotrowski, John D. "Alignment Techniques." Proceedings, Machinery Vibration Monitoring and Analysis, New Orleans, June 1984, pp. 214-246.

7
Measuring Thermal Movement of Rotating Machinery

Virtually all rotating equipment will undergo a change in position during start up and while running that affects the alignment of the shafts. In order for the shafts to run colinear under normal operating conditions, it is desirable to know the amount and direction of this movement to position the machinery properly during the "cold" alignment process to compensate for this change. In many cases, this movement is negligible. In other cases it can make all the difference in the world between a smooth-running drive system and one that is not. It is important to know what sort of movement is occurring before you ignore it.

There are a variety of factors that cause machinery to move once it is on line and running. The most common cause is due to temperature changes in the machinery itself (as it compresses gases or heats the lubricant from friction in the bearings) and is therefore generally referred to as "thermal" movement. The temperature change in rotating machinery is rarely uniform throughout the casing, which causes most equipment to "pitch" at some angle rather than grow (or shrink) straight up (or down). For compressors and pumps, thermal movement of the attached piping will also cause the equipment to shift.

Since there is a possibility that the equipment may move when any rigid attachments are connected, perform the cold shaft alignment process before attaching any piping (on pumps, compressors, or steam turbines) or rigid conduit (on motors and generators), then watch what happens to the position of the shafts after all the auxiliary hardware has been put on. No movement should occur.

Other sources of movement in machinery can be caused by settling of foundations, loose or cracked foot bolts, varying weather conditions for equipment located outdoors, heating or cooling of concrete pedestals, changes in the operating condition of equipment from unloaded to loaded postures, and casing and support counterreactions to the centrifugal force of rotors as they are rotating.

Some special considerations must be taken into account for equipment that is started and stopped frequently or where loads may vary considerably while running. In cases like these, a compromise has to be made that weighs factors such as period of time at certain conditions, total variation of machinery movement from maximum to minimum, coupling and alignment tolerances, etc. To observe and record these changes properly, periodic checks should be made of this change in movement to understand how to position the equipment effectively for optimum performance. Continuous monitoring systems are available and are explained later in this chapter. It has been my experience, however, that the majority of rotating equipment will maintain one specific position regardless of varying loads. What usually turns out to be a bigger problem is that some equipment may have to be offset aligned "cold" by a considerable distance, making start ups very nerve wracking. In most cases, equipment will undergo the greatest rate of change of movement shortly after start up. "Shortly" can mean anywhere from 5 minutes to 1 hour for most types of equipment and may settle at some position hours or even days later. Consequently the reverse is true; equipment will experience almost an equivalent amount of movement right after it has been shut down.

For many years people have attempted to take quick "hot" alignment readings, that is, take dial indicator shaft alignment readings immediately after a unit has been shut down. This turns out to be quite a difficult task to perform for a variety of reasons. Mounting brackets and indicators quickly enough to get a set of readings, safety tagging driver units to prevent them from inadvertently starting up with the brackets mounted on the shafts and getting an accurate set of readings while the shafts are still moving, proves to be a real challenge for the personnel doing the work. It is wise to take three or four readings (say, every five minutes) during the "cool" down period to plot the movement and then extrapolate them back to the instant when the unit was first shut down to determine the actual shaft positions when running. The data are usually nonlinear, and guessing the slope of the curve during the first five-minute period is a hit-or-miss proposition.

There are many inventive ways to measure shaft alignment positions from cold to hot and this chapter will review the more commonly used techniques. Each method has its advantages and disadvantages and it is a good idea to compare the results from two or more methods, just to see if the results are similar.

MATHEMATICAL SOLUTION

At the atomic level in solid materials, the temperature and volume of the material is dictated by the vibration of the individual molecules. In other words, the hotter a solid material gets, the more the molecules vibrate and the farther apart the molecules are spaced. This phenomenon causes changes in dimensions (i.e., strain) that can be calculated by the following equation:

$$\Delta L = L\ \alpha\ \Delta T \tag{7.1}$$

where:

- ΔL = change in dimension that occurs from the change in temperature of the material (in.)
- L = length of the object (in.)
- α = coefficient of thermal expansion (or contraction) (in./ in.-°F)
- ΔT = change in temperature (°F)

The coefficients of thermal expansion for the majority of materials used in machinery casing and foundation are shown in Table 7.1. These coefficients can be used for temperatures

TABLE 7.1 Coefficients of Thermal Expansion for
Different Materials

Material	α (in./in.-°F)
Aluminum (99% pure)	12.0×10^{-6}
Aluminum alloys	12.5×10^{-6}
Brass (70 Cu - 30 Zn)	11.0×10^{-6}
Carbon steel (1040)	6.3×10^{-6}
Cast iron (grey)	5.9×10^{-6}
Concrete	$6.5 - 8.0 \times 10^{-6}$
Nickel steel	7.3×10^{-6}
Stainless steel	9.6×10^{-6}
Vulcanized rubber	45.0×10^{-6}
Nylon	55.0×10^{-6}

between 32–212°F. There is a slight shift in the value of the coefficient for higher or lower temperatures due to the nonlinearity
of molecular vibration in materials.

Sample Problem 1

Figure 7.1 shows a centerline mounted barrel-type compressor
with carbon steel stands, 36 in. in height. When the compressor is not running, the surface temperature of the stand is 70°F.
After the compressor has been operating long enough to achieve
thermal stability, the temperature of the stands is measured to
be 115°F (average). The change in temperature of the stands
from cold to hot is therefore 45°F. The change in height can be
calculated as follows:

$\Delta L = L \, \alpha \, \Delta T$
$\Delta L = (36 \text{ in.}) \times (6.3 \times 10^{-6} \text{ in./in.-°F}) \times (45°F)$

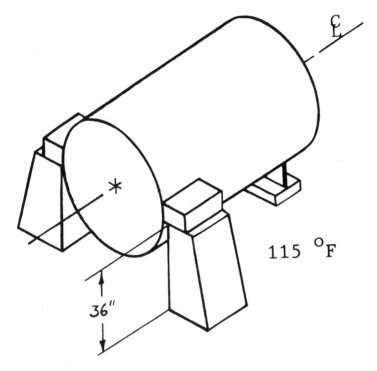

FIGURE 7.1 Compressor stand height and temperature change.

$\Delta L = 0.00972$ in.
$\Delta L \cong 0.010$ in.

This 10 mil growth of the stand will obviously affect the position of the compressor centerline at the point where the stand is attached. Always measure the change in height at the other end of the unit to determine what value it may change to see if the unit is undergoing a rise straight up or if it is pitching at some angle.

Sample Problem 2

Figure 7.2 shows the other end of the same compressor where the sway bar is attached. The change in height is slightly more

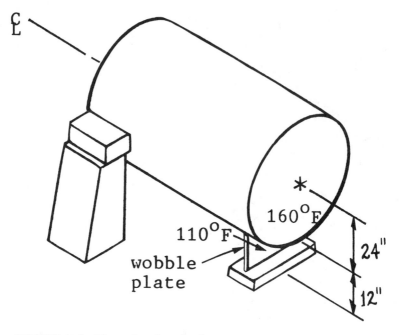

FIGURE 7.2 "Sway bar" end of compressor.

difficult to calculate since there is a difference in the temperature of the compressor casing and the temperature of the sway bar.

Compressor Case

$\Delta L = L \, \alpha \, \Delta T$
$\Delta L = (24 \text{ in.}) \times (6.3 \times 10^{-6} \text{ in./in.-°F}) \times (90°F)$
$\Delta L = 0.0136 \text{ in.}$
$\Delta L \cong 0.014 \text{ in.}$

Sway Bar

$\Delta L = L \; \alpha \; \Delta \; T$
$\Delta L = (12 \text{ in.}) \; X \; (6.3 \; X \; 10^{-6} \text{ in./in.-}°F) \; X \; (40°F)$
$\Delta L = 0.0030 \text{ in.}$

Total rise = 0.014 in. + 0.0030 in. = 0.017 in.

INFRARED PHOTOGRAPHIC TECHNIQUES TO DETERMINE THERMAL PROFILES OF ROTATING EQUIPMENT

Turbomachinery casings transfer heat to the environment when running that is generated from friction, compression of process gases, steam flow, or motor stator windings. Surface temperatures at various locations on a casing vary widely, and virtually no rotating equipment maintains a constant thermal gradient across the entire casing. Gas turbines, for instance, may have inlet casing temperatures below the ambient air temperature, and six feet away have a 1200°F casing temperature at the combustor section. Since our eyesight is limited only to the visible spectrum (400-700 nm), we are unable to see the temperature gradient profile of machine cases as they emit the longer, infrared radiation. Infrared radiation can be categorized in four general ranges:

Actinic range: incandescent objects such as bulb filaments having wavelengths near the visible red region
Hot object range: objects with temperatures around 400°F
Calorific range: objects with temperatures around 250°F
Warm range: objects having temperatures below 200°F (approximately 9000 nm)

The infrared radiation emitted from an object can be recorded either on infrared sensitive film or by electronic thermography equipment. The type of equipment used for thermographic studies is shown in Figure 7.3. These instruments scan the object

FIGURE 7.3 Thermographic scanning equipment. Photo courtesy of AGA Corp., Secaucus, N.J.

for the infrared radiation and amplify the converted electrical signals from a supercooled photodetector onto a cathode ray tube (CRT), where a photographic image of the object can be recorded.

Figure 7.4 shows a three-stage centrifugal compressor case, and Figure 7.5 illustrates the temperature profile when the compressor is running under full load. The white areas show where the infrared radiation (heat) is the greatest. The hottest areas in this image are approximately 135°F.

Figure 7.6 shows an axial flow compressor with rigid supports at the inlet end and flexible supports at the discharge end. Figure 7.7 illustrates the thermal profile of the discharge end with the compressor running under load (note the hot spot at the 1 o'clock position). Figure 7.8 shows a closer view of

FIGURE 7.4 Compressor case. Photo courtesy of General Electric Co., Cincinnati, Oh.

the flexible support leg. The lifting eye is at the left side of the photograph and the flex leg is the black portion just to the right of the lifting eye. The photograph clearly shows that the support leg stays at ambient temperatures and does not expand thermally (as originally thought when the machinery was installed).

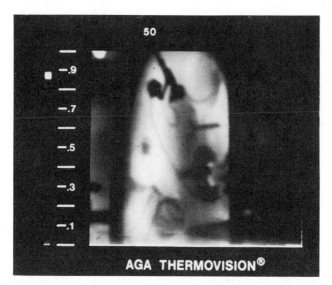

FIGURE 7.5 Thermal image of compressor case. Photo courtesy of General Electric Co., Cincinnati, Oh.

FIGURE 7.6 Axial flow compressors. Photo courtesy of General Electric Co., Cincinnati, Oh.

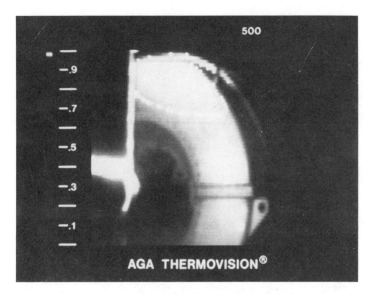

FIGURE 7.7 Thermal image of compressor and casing. Photo courtesy of General Electric Co., Cincinnati, Oh.

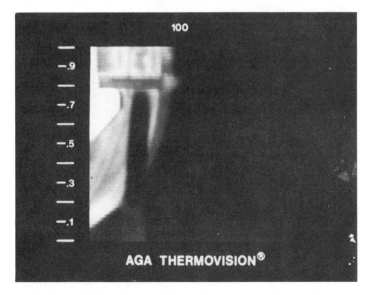

FIGURE 7.8 Thermal image of the support leg. Photo courtesy of General Electric Co., Cincinnati, Oh.

FIGURE 7.9 Induction motor. Photo courtesy of General Electric Co., Cincinnati, Oh.

Figure 7.9 shows an end view of a 9000-hp induction motor, Figure 7.10, a close up of the bearing area, and Figure 7.11, the lower corner at the axial jackscrew. Note that the heat generated in the bearing does not conduct through the entire end frame of the motor.

Figure 7.12 shows an aircraft derivative gas and power turbine driver set. Figure 7.13 illustrates the thermal image when

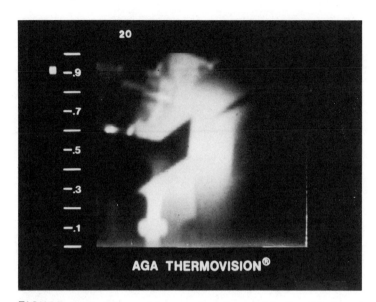

FIGURE 7.10 Thermal image of motor end bell. Photo courtesy of General Electric Co., Cincinnati, Oh.

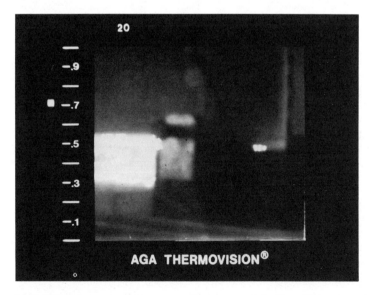

FIGURE 7.11 Lower corner of motor end bell. Photo courtesy of General Electric Co., Cincinnati, Oh.

FIGURE 7.12 LM 1500 gas and power turbine assembly. Photo
courtesy of General Electric Co. , Cincinnati, Oh.

running. The white area in the thermal photograph is the combustor section of the gas turbine.

Although movement of rotating machinery casings do not occur solely from temperature changes in the supporting structures
and the casings themselves, infrared thermographic studies can
assist in understanding the nature of the thermal expansion taking
place.

USING INSIDE MICROMETERS AND A
BEVEL PROTRACTOR TO MEASURE
MACHINERY MOVEMENT

One of the easiest and least expensive techniques for measuring
casing movement in the field can be done with inside micrometers

FIGURE 7.13 Thermal image of gas turbine and power turbine. Photo courtesy of General Electric Co., Cincinnati, Oh.

and a bevel protractor (shown in Figure 7.14). Tooling balls or similar reference point devices are rigidly attached to the foundation and the machine casing. Readings between the tooling balls are then taken when the machinery is at rest and taken again when the equipment is running and has stabilized thermally, as shown in Figure 7.15. The distances and angles (see Figure 7.16) may then be plotted to show the resulting horizontal and vertical movement of the machine casing at each bearing as shown in Figure 7.17.

OPTICAL TOOLING EQUIPMENT AND TECHNIQUES

Jig transits, optical tooling levels, and lasers are probably the most versatile equipment available to determine rotating equipment movement. Figures 7.18 and 7.19 show the two most widely used optical instruments for machinery alignment. This chapter will deal specifically with their ability to measure this movement

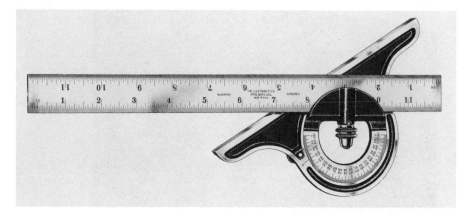

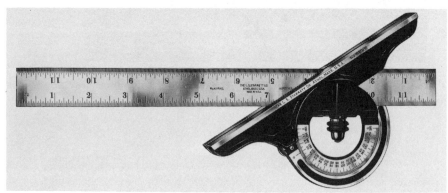

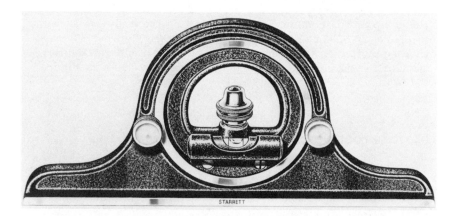

FIGURE 7.14   Bevel protractors (above) and inside micrometers
(facing page). Courtesy of L. S. Starett Co., Athol, Mass.

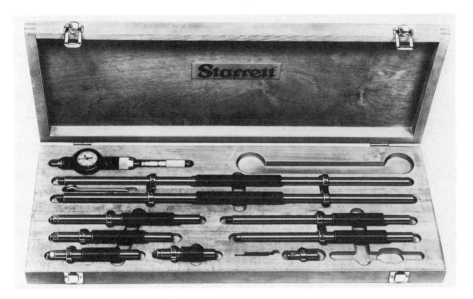

FIGURE 7.14 (continued)

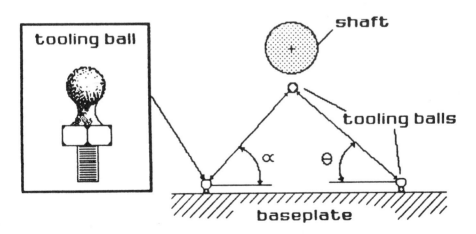

FIGURE 7.15 Tooling balls and location spots on machine casing.

$$a^2 = b^2 + c^2 - 2bc \cos \alpha$$

$$\alpha = \cos^{-1}\left[\frac{a^2 + c^2 - b^2}{2ac}\right]$$

$$\theta = \sin^{-1}\left[\frac{b \sin \alpha}{a}\right]$$

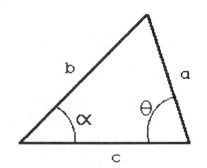

FIGURE 7.16 Laws for oblique triangles.

but will not even begin to explain their full potential for many other uses, such as leveling foundations and checking dimensions on three- or four-bearing machines. For companies that have a large quantity of rotating equipment or are undergoing a major expansion, it is highly recommended that this equipment be

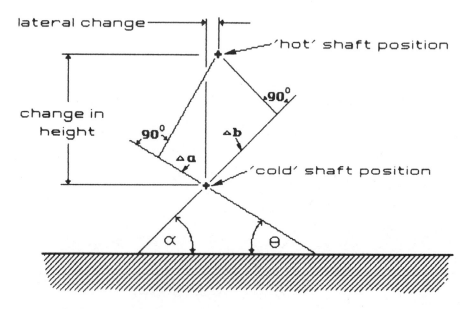

FIGURE 7.17 Movement plot.

FIGURE 7.18 Optical tooling level.   Courtesy of Keuffel & Esser
Company, Morristown, N.J.

purchased and that someone be trained in its use. One major manufacturer offers an excellent two-week school on optical tooling,
emphasizing hands-on training. If you're not interested in doing
this yourself, there are some consulting firms that specialize in
using optical alignment instruments for industrial applications.

The scale targets used for optical tooling are shown in Figure 7.20. There are generally four sets of sighting marks on
the scales for centering of the crosshairs. An optical micrometer is attached to the instrument and can be positioned in either
the horizontal or vertical direction. The adjustment wheel is
used to align the crosshairs between the scale sighting lines on
the targets. When the crosshair is lined up between the sighting lines, a reading is taken based on where the crosshair is
sighted on the scale and the position of the optical micrometer
as shown in Figure 7.21.

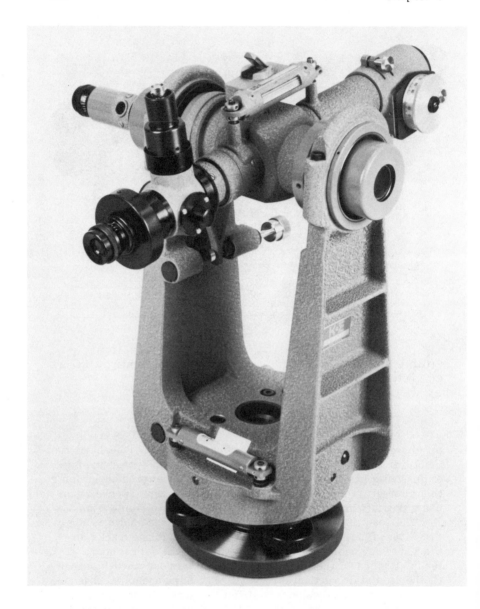

FIGURE 7.19 Jig transit. Courtesy of Keuffel & Esser Company,
Morristown, N.J.

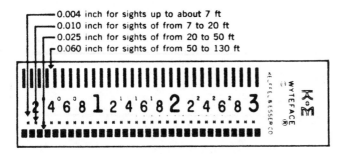

FIGURE 7.20 Typical scale target showing standard target pattern of each tenth. Courtesy of Keuffel & Esser Company, Morristown, N.J.

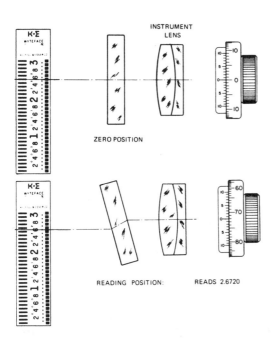

FIGURE 7.21 Principle of an optical micrometer. Courtesy of Keuffel & Esser Company, Morristown, N.J.

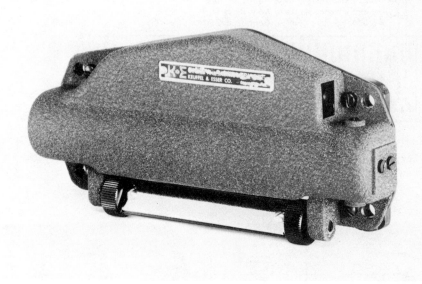

FIGURE 7.22   The coincidence level (shown above) operates accor-
ding to the principle diagrammed on the facing page. Courtesy of
Keuffel & Esser Company, Morristown, N.J.

     The extreme accuracy (one part in 200,000 or 0.001 in. at a
distance of 200 in.) of the optical instrument is obtained by a
split coincidence level mounted on the telescope, as shown in
Figure 7.22.
     To level the instrument correctly, perform the following pro-
cedure:

     1. Set the instrument stand at the desired sighting location,
attach the alignment scope to the tripod or instrument stand, and
level the stand using the circular bubble level on the tripod. Insure that the stand is steady and away from heat sources, vibrating floors, and curious people who may want to use the scope
to see sunspots.
     2. Rotate the scope barrel to line up with two of the four
leveling screws and adjust these two leveling screws to center

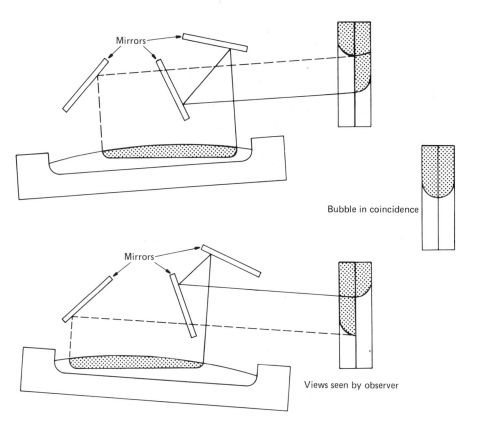

Bubble in coincidence

Views seen by observer

FIGURE 7.22 (continued)

the bubble roughly in the same tilt plane as the two screws that are being adjusted (Fig. 7.23).

3. Rotate the scope barrel 90° to line up with the other two leveling screws to center the bubble completely in the circular level (Fig. 7.24).

4. If the circular level is still not centered, repeat Steps 2 and 3.

5. Once again, rotate the scope to line up with two of the leveling screws, as shown in Figure 7.23. Adjust the tilting screw to center the split coincidence level on the side of the scope barrel (Fig. 7.25).

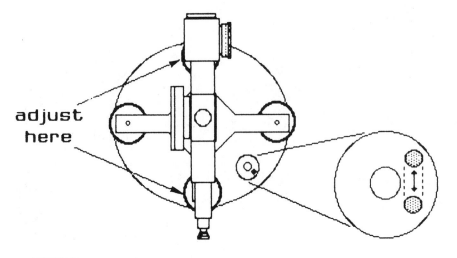

FIGURE 7.23 Adjusting the tilt screws in one axis.

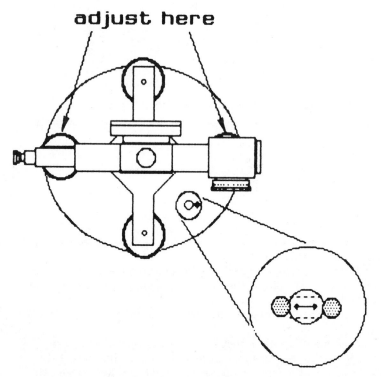

FIGURE 7.24 Adjusting the tilt screws to center the bubble in the circular level.

bubble in coincidence

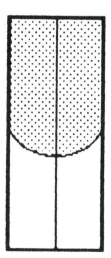

FIGURE 7.25 Adjust the tilting screw to center the split coincidence bubble. Photo courtesy of Keuffel & Esser Company, Morristown, N.J.

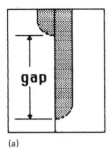
(a)

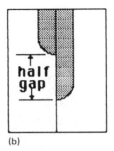
(b)

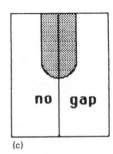
(c)

FIGURE 7.26 Position of the split coincidence level before and after adjustments: (a) before level screw adjustment; (b) after level screw adjustment; (c) after tilting screw adjustment.

6. Rotate the scope barrel 180° and note the position of the two bubble halves. Adjust the two leveling screws in line with the scope barrel so that the gap between the two bubble halves is exactly one half the original gap (Fig. 7.26a, b, c).

At this point, adjust the tilting screw so there is no gap in the two bubble halves. Rotate the scope barrel back 180° to its original position and see if the two bubble halves are still coincident (i.e., there is no gap). If they are not, adjust the two leveling screws and the tilting level screw again as shown in Figures 7.23 and 7.25 and rotate the scope barrel back 180° until there is no gap when swinging back and forth through the half-circle. At this point, the two leveling screws should be snug but not tight enough to warp the mounting frame.

7. The last step is to rotate the scope barrel 90° to line up with the two remaining leveling screws yet to be fine adjusted. Follow the same procedure as outlined in Step 6 and Figure 7.24. When these adjustments have been completed, the split coincidence bubble should be coincident when rotating the scope barrel through the entire 360° of rotation around its azimuth axis.

If there is any change in the split bubble gap during the final check, go back and perform this level adjustment again. This might take a half-hour to an hour to do correctly, but it is time well spent. It is also wise to walk away from the scope for about 30 minutes to determine if the location of the instrument is stable and to allow some time for your eyes to uncross. If the split coincident bubble has shifted during your absence, start looking for problems with the stand or what it is sitting on. Correct the problems and relevel the scope.

I cannot overemphasize the delicacy of this operation and this equipment. It is no place for people in a big hurry with little patience. If you take your time and are careful and attentive when obtaining your readings, the accuracy of the equipment will astonish you.

Optical Parallax

As opposed to binoculars, 35-mm cameras, and microscopes that have one focusing adjustment, the optical scope has two focusing knobs. There is one knob used for obtaining a clear, sharp image of an object (the target scale) and another adjustment knob that is used to focus the crosshairs (the reticle pattern). Since your eye can also change focus, it is important to adjust both these knobs properly so that your eye is relaxed when the object image and the superimposed crosshair image is focused on a target.

Adjusting the Focusing Knobs

1. With your eye relaxed, aim at a plain white object at the same distance away as your scale target and adjust the eyepiece until the crosshair image is sharp.
2. Aim at a scale target and adjust the focus of the telescope.
3. Move your eye slightly sideways and then up and down to see if there is an apparent motion between the crosshairs and the target you are sighting. If so, defocus the telescope and adjust the eyepiece to refocus the object. Continue alternately adjusting the telescope focus and the eyepiece to eliminate this apparent motion.

Checking the Calibration of the Instrument – The Peg Test

Before using any optical instrument, it is imperative that a calibration check be performed to insure instrument accuracy.

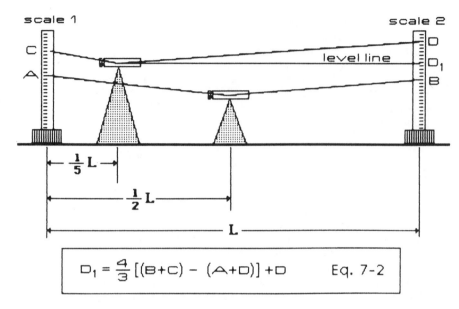

$$D_1 = \frac{4}{3}\left[(B+C) - (A+D)\right] + D \qquad Eq. \ 7\text{-}2$$

FIGURE 7.27 Peg test setup.

The Peg Test

1. Set two scales apart by distance L (usually 40 ft) and the telescope exactly halfway between both scales. Use a plumb bob attached to the instrument stand in line with the azimuth axis to aid in positioning the instrument exactly at the halfway point. Accurately level the instrument using the split coincident level.

2. Alternately take four readings on scale 1 (reading A) and scale 2 (reading B). Record and average these readings.

3. Move the scope to the 1/5 L position, level the scope, and alternately take four readings on scale 1 (reading C) and scale 2 (reading D). Record and average these readings.

4. If the instrument is calibrated, A minus B should equal C minus D. If this is not the case, the coincident level calibration adjustment nuts can be adjusted to position the leveled line of sight to be set at reading D1 (see Fig. 7.27). Consult the manufacturer of the scope to assist in making this adjustment. Be extremely cautious when doing this to prevent damage to the

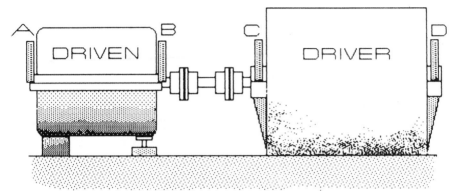

FIGURE 7.28 Side view of optical scale targets positioned on each piece of machinery near the centerline of rotation for use with the optical tilting level.

instrument and coincident leveling system. The manufacturer will usually be happy to make this adjustment for you for a small fee should you decide that it is beyond your ability. *Note*: At 40 ft, the accuracy of the scope is plus or minus 0.0024 in.

Using Optical Tooling for Measuring Machinery Movement

General Procedure

1. Check the calibration of the instrument.
2. Select the scale positions on the machinery at (or as close as possible to) the centerline of rotation and usually near the bearings or the outboard feet of each piece of rotating equipment in the drive train.
3. Set the optical instrument and stand at some remote reference point away from the drive train where a stable point in space can be established, but close enough to maintain the maximum accuracy of the readings.
4. Level the instrument accurately and take a set of readings at each target scale mounted on the machinery when it is cold (i.e., not running). See Figures 7.28 and 7.29 for possible scale arrangements.

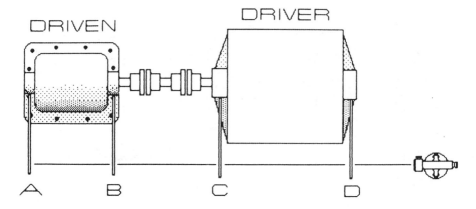

FIGURE 7.29 Top view of scale targets set to monitor sideways movement of the machinery using a jig transit.

5. Run the machinery at normal conditions and allow the equipment to stabilize thermally.
6. Check the level accuracy and take a similar set of readings at each target scale.
7. Compare the cold set of readings to the hot set of readings to determine the amount and direction of the movement of each scale (see Table 7.2).
8. Plot this movement data on a graph (as shown in the example in Figure 7.30) or use this information in the mathematical calculations for determining the proper equipment moves as outlined in Chapter 6.

TABLE 7.2 Example of Data Collected on Scale Targets to Determine Vertical Movement

Scale	A	B	C	D
Cold	7.652	6.987	4.232	5.010
Hot	-7.648	-6.976	-4.207	-4.996
Change	+0.006	+0.011	+0.025	+0.014

Note: + up; - down.

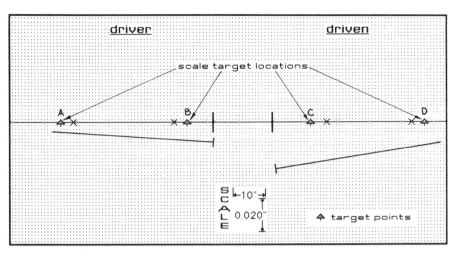

FIGURE 7.30 Plotting the data from Table 7.2 onto graph paper to show the desired "cold" position of the two equipment shafts.

LASER ALIGNMENT EQUIPMENT
AND TECHNIQUES

The basic principles of laser alignment techniques are strikingly similar to those of the optical alignment method previously described. Instead of establishing a level line of sight as in optical equipment, the laser projects a straight beam of coherent light at a photosensitive target located on the machinery that can measure the change of position of the striking light beam as the target and machinery move. The light detector target is coupled to a readout device that displays the displacement of the laser beam from target center both vertically and laterally in 0.001-in. increments. Figure 7.31 shows the components of the system.

Laser: low power helium neon laser. Laser beam is projected
   through an optical lens system to produce an effective beam
   exit diameter of 3/8 in. At 300 ft, the beam diameter is
   11/16 in. Can be mounted on a variety of stands.
Power supply: provides a 1200 VDC/5 ma source to drive the laser. The DC signal is AC modulated at 10 Khz.

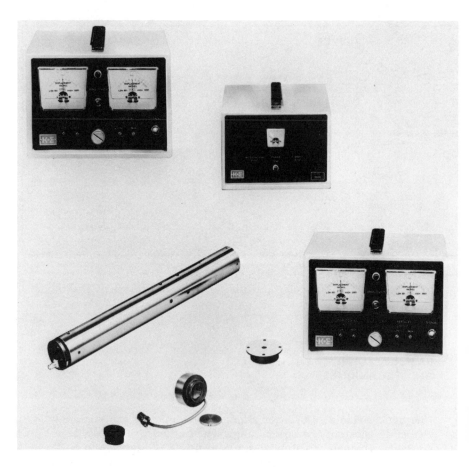

FIGURE 7.31 Laser alignment hardware. Photo courtesy of
Keuffel & Esser Company, Morristown, N.J.

Readout unit: receives and conditions the signal from the quad
     cell detector targets (up to four target inputs) and displays
     deviation of laser beam center from target center in 1-mil in-
     crements in both vertical and horizontal directions. When the
     laser beam is up and to the left of target center, the displayed
     values are positive. When the laser beam is down and to the

right, the displayed value is negative. Range is plus or minus 0.100 in. Can be connected to a hard copy printer.
Quad cell detector target: senses change of striking laser beam
center from target center.

The laser instrument could be used with the universal stand,
optical micrometer, and split coincident level, as shown in Figure
7.32.

Using Laser Tooling for Measuring Machinery Movement

General Procedure

1. Decide on the location of laser and stand, check calibration of coincident level if used, and attach miscellaneous equipment as desired. The laser becomes the fixed reference point,
so be sure the location is stable, vibration free, and away from
the machinery baseplate and foundation if possible.
2. Select locations for placement of the detector targets as
close to the bearing centerline as possible. Provide some sort of
positioning system that can orient the detector target in the vertical and lateral directions.
3. Turn on the laser system as outlined in the instructions
and allow a one-hour warm up period to allow the laser system
to stabilize.
4. Place a dust cover over the detector target to begin rough
aligning the quad cell target to center the laser beam. Remove
the dust cover after rough aligning and begin the trim adjustments by observing the displays on the readout unit until the
beam has been centered in both the horizontal and vertical directions. If see through quad cells are used (Fig. 7.34; see
Figure 7.35 for a picture of a quad cell), start with the near-
est target to the laser and continue adjusting each target along
the drive train so all are centered with the drive train not running. If the quad cells are arranged as shown in Figure 7.33,
direct the laser beam at each target insuring that the laser is
properly leveled with the attached split coincidence bubble.
When using this type of arrangement, the laser is leveled the
same way as an optical telescope that was outlined in Figures
7.23 through 7.26. In this setup, remember that the horizontal

(a)

FIGURE 7.32 Optional equipment for laser system. (a) universal stand; (b) optical micrometer; (c) coincidence level. Photos
courtesy of Keuffel & Esser Company, Morristown, N.J.

readings are meaningless, since you won't know the exact position of the laser beam as it is rotated through its azimuth axis
when aiming at each successive quad cell target along the drive
train. Therefore, during the cold and hot readings, position
the laser beam so that the horizontal readings are zeroed each
time a new target is sighted. *Note*: If the drive train has a
preheated, pressure-fed lubrication system, insure that the
lube system is not running when the detector targets are being
centered. The detector targets should be zeroed with the drive
train in the same condition as if the machinery shafts are being
aligned with dial indicators, as explained in Chapter 5.

(b)

(c)

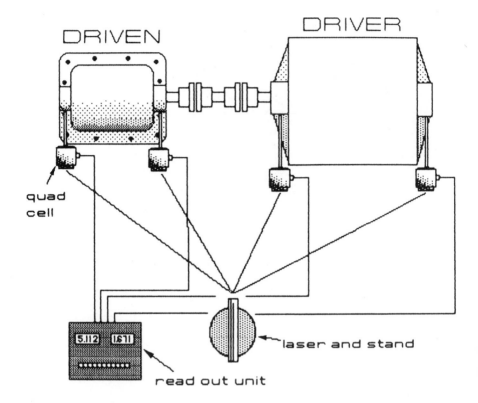

FIGURE 7.33 Typical laser and quad cell detector target positions for vertical movement only.

    5. Start up the drive train and allow the machinery to stabil-
ize at normal operating conditions. Log the readings on the readout unit for every quad cell target.
    6. This entire procedure should be performed at least twice
to determine if the machinery movement is repeatable. If the deviation in movement of the readings on the machine elements fluc-
tuates more than ±10%, begin looking for problems with the laser
stand, the target brackets, or with the machinery foundation or
piping arrangement itself.

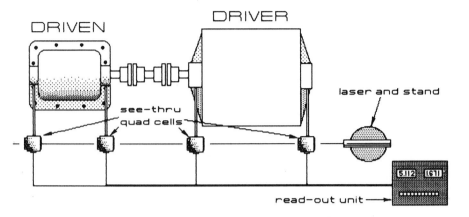

FIGURE 7.34 Laser and detector target positions for both vertical and horizontal movement using see-through quad cell targets.

MONITORING MACHINERY MOVEMENT
WITH PROXIMITY PROBES

Since the advent of the eddy current proximity probe in the 1960s,
many ingenious applications have been found for its use in industry besides monitoring shaft vibration or thrust position of rotors.
Since the proximity probe is basically an electronic dial indicator,
it is ideally suited for measuring changes in gap between the
probe tip and a conductive surface such as carbon steel or cast
iron. There are two popular prox probe mounting techniques
employed to observe movement of rotating machinery: (1) water
cooled pipe stands; and (2) Dodd alignment bars. The basic
principles of proximity probes are illustrated in Figure 7.36.

Water-Cooled Pipe Stands

This arrangement is illustrated in Figure 7.37. The proximity
probes are attached with a bracket to a water-cooled pipe stand
which is firmly anchored to the machinery foundation near each

FIGURE 7.35 "See through" detector target. Courtesy of Keuffel
& Esser Company, Morristown, N.J.

bearing. To maintain a stable reference point, water should be
circulated through the pipe stand or the pipe should be insulated
and filled with a water-glycol solution to prevent as little dimensional change as possible that could occur to the pipe stand itself

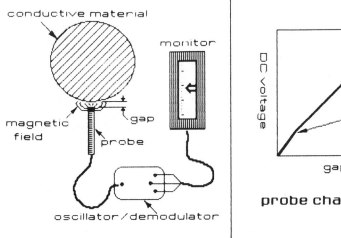

FIGURE 7.36 Principle of proximity probes.

monitoring the shaft directly

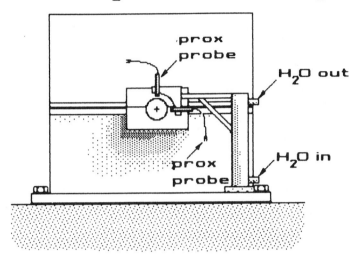

monitoring a steel block near the shaft

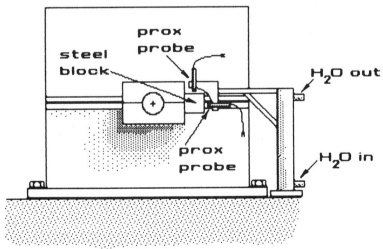

FIGURE 7.37 Typical pipe stand arrangement at one end of a machine casing monitoring the position of the shaft.

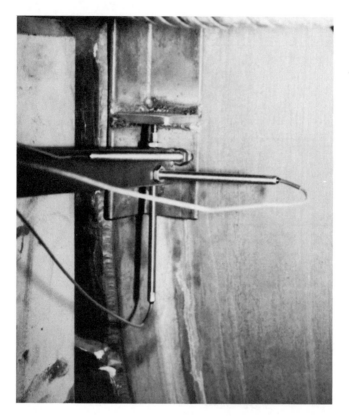

FIGURE 7.38 Probes mounted in three axes.

from radiant heat emitted from the machinery. The probes are mounted to a bracket on the pipe stand and positioned to monitor an L-shaped target affixed to each end of a machine casing, as shown in Figures 7.38 and 7.39. Movement can be monitored in the horizontal, vertical, and axial directions as shown in these figures.

The probes could also be positioned to monitor the movement of the shaft directly, since it is really the position of the shaft that is to be determined from a cold to a hot position, as shown in Figure 7.39. It is important to insure that the probe tips are far enough apart to prevent any cross-field effects from one

FIGURE 7.39 Pipe stand and bracket arrangement.

probe to another that will affect accurate gap measurements. The
probes should always be statically calibrated to the same type of
material being observed, since the gap versus voltage characteristics are different from one material to another. The probe manufacturer can supply information on these curves and the procedure for calibrating probes for different materials. If the direction of machinery movement is not known when the probes are initially gapped, some adjustments may be necessary after the first
attempt in case the target (or shaft) is moving too close or too
far away from the probe tip to keep the probe within its linear
range. Standard probes are usually good for gap changes near
80 mils, and some manufacturers can supply special probes able
to measure up to 1/2 in. of gap change. Once again, these pipe
stands must be mounted at both ends of each machine case to determine actual machine movement for alignment purposes. The

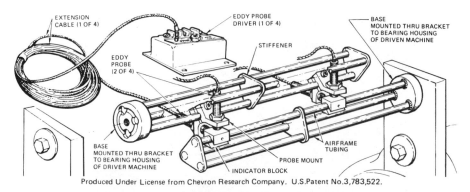

Produced Under License from Chevron Research Company. U.S.Patent No.3,783,522.

FIGURE 7.40 Alignment bar system. Diagram courtesy of Scientific Atlanta, Atlanta, Ga.

only disadvantage to this system is that it cannot measure any change in the machinery foundation itself, and it is quite possible that a change could occur.

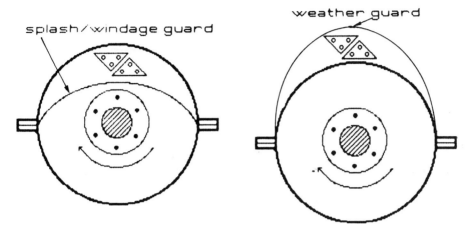

FIGURE 7.41   Windage guards.

MACHINERY ALIGNMENT BARS

Stated simply, machinery alignment bars are electronic reverse
dial indicators. One bar is attached near the centerline of rotation on one machine element near the coupling with two probes
attached to the end of the bar that look at a steel block attached
to another bar mounted to the other machine element, itself having
two probes attached to its bar end (Fig. 7.40). Before starting
the drive train, a set of gap readings are taken on each probe
and logged. The drive system is then run until the probe gaps
have stabilized during normal operating conditions. Relative machinery casing movement can be determined by comparing the
gaps on each probe before and after the equipment was running.
Strip chart recorders can be set up to monitor the rate of change
of gap during warm-up and on-line operating conditions.

Since the bracket will be located near the coupling, some special precautions are warranted to prevent movement of the bracket
and probe arrangement from coupling windage or splashing oil,
as shown in Figure 7.41. Even though the overhung brackets
may sag when attached to a machinery case, the sag should re-
main constant from cold to hot machinery conditions. Slight
amounts (up to 5 mils, as a rule of thumb) of bar end vibration

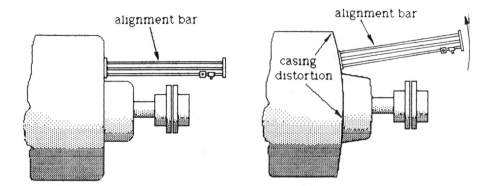

FIGURE 7.42 Machinery casing distortion where alignment bars
are mounted.

can be discounted, since the proximity probe system is averaging
the gap from any oscillatory motion occurring on the target.

For bracket spans longer than 10 in., dry instrument air can
be blown down through the center of each of the three bracket
support tubes to limit the amount of thermal expansion or contraction.

Another problem could occur if the brackets are mounted improperly on the machinery case due to casing distortions from
unequal thermal profiles across the machinery, as shown in Figure 7.42.

Taking infrared thermography surveys of the machine casing
area where the brackets will be attached will be helpful in determining the best location for the brackets.

If the alignment bar system is going to be used on a number
of different drive systems with varying distances between machine casings, it may be financially advantageous to have bracket
extensions of various lengths that can be bolted end-to-end to
span the various distances.

Offset Aligning Rotating Machinery to
Compensate for Thermal Movement

Once accurate measurements have been collected and analyzed on
how the equipment moves in the field, the machine elements can

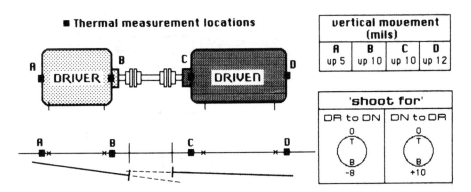

FIGURE 7.43 Typical thermal movement data.

then be properly positioned during the dial indicator alignment
process to compensate for this movement and to insure colinear
shaft centerlines during operating conditions. To obtain the
necessary cold shaft positions, the thermal movement data can
be plotted graphically, as shown in Figure 7.43.

The distances between the thermal measurement points are
taken and placed axially along the drive train with the other critical items, such as the dial indicator locations on the coupling
hubs of shafts, foundation bolts, etc. Figure 7.43 illustrates
how this is set up for optical, laser, inside micrometers/bevel
protractors, and water-cooled proximity probe pipe stands.
Once the graph is laid out, "shoot-for" reverse indicator or face
peripheral dial indicator readings can be determined for the shaft
positions when at rest, using the graphical alignment calculator
from Chapter 6. Remember that the dial indicator reading will be
twice the amount indicated by the chart.

If the machinery alignment bar system was used to determine
the machinery movement, the graph setup would look like Figure 7.44. A little bit of thought is going to have to be put into
recalling how the probes were positioned when reading the targets and what decreasing or increasing gaps mean when setting
up the chart. It is too easy to make a mistake here by misinter-
preting the movement data, so it is wise to make sure both the
amount of movement and the direction of movement is correct and
that you have gone over the graph setup at least twice before
running out and positioning the machinery with "shoot for" readings that are incorrect.

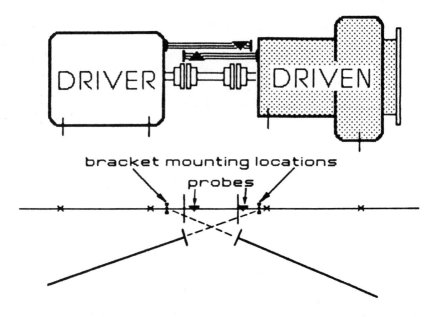

FIGURE 7.44 Thermal movement plot using alignment bar systems.

If alignment computers or mathematical calculation techniques are used to determine the correct shim changes or lateral movements, apply the same geometric principles explained in Chapter 6.

Determining the "Shoot For" Cold
Reverse Indicator Readings

Apply these general procedures to determine what the "shoot for" readings will be when offset-aligning machinery to compensate for dynamic machinery movement.

1. Graph the desired cold offset shaft positions of both the driver and driven units. Figure 7.45 shows two units plotted in both the vertical and horizontal planes. In the vertical plane,

DRIVER   DRIVEN

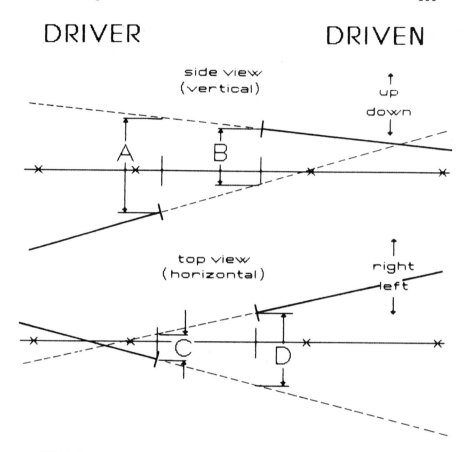

FIGURE 7.45 Plotted thermal movement in both the horizontal and vertical planes.

the driver centerline is positioned below the chart centerline to indicate that it rises when going from the cold to hot position. The amount of movement of this centerline is based on the data collected from any of the thermal measurement techniques explained in this chapter. The driven unit centerline is positioned above the chart centerline to indicate that it lowers when going from a cold to hot position. In the horizontal plane, the outboard end of the driver unit moves to the right and the inboard end moves to

driver to driven

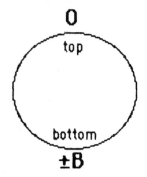

driven to driver

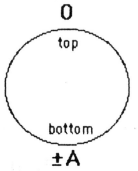

FIGURE 7.46 Recording the vertical distance between the two shafts where the dial indicator stems will be positioned during alignment.

the left. The driven unit moves to the right at both the inboard and outboard ends but different amounts at each end.

2. Based on the chosen scale factor from top to bottom on the chart, record the A and B gaps for the vertical position, as shown in Figure 7.45, and log in a manner similar to Figure 7.46.

3. Determine whether the bottom readings are positive or negative on the coupling.

RULE

If the actual centerline of a unit is toward the bottom of the graph with respect to a projected centerline, the reading will be positive. If the actual centerline of a unit is toward the top of the graph with respect to a projected centerline, the reading will be negative.

In other words, try to visualize what is going to happen to the dial indicator stem as it traverses circumferentially from top to bottom on the shaft or coupling hub of each machine. Is it going to move outward (negative) or inward (positive)? Figure 7.45 shows that the driver centerline appears to be lower from the vantage point of the driven unit, therefore the dial indicator stem will move inward as it rotates to the bottom of the driver coupling hub, producing a positive reading. From the vantage

driver to driven

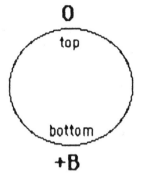

driven to driver

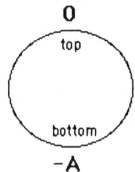

FIGURE 7.47 Determining the appropriate sign for the shafts shown in Figure 7.45 for vertical alignment.

point of the driver unit however, the dial indicator stem will move outward as it rotates to the bottom of the driven coupling hub, producing a negative reading as shown in Figure 7.47.

4. Based on the chosen scale factor from top to bottom on the chart, record the C and D gaps (shown in Figure 7.45) for the horizontal view (shown in Figure 7.48).

driver to driven

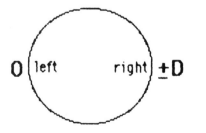

driven to driver

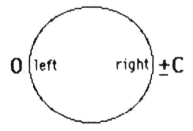

FIGURE 7.48 Recording the horizontal distance between the two shafts where the dial indicator stems will be positioned during alignment.

driver to driven   driven to driver

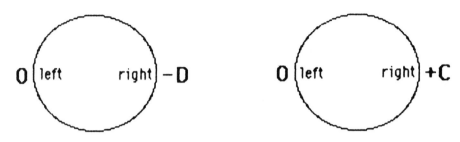

FIGURE 7.49 Determining the appropriate sign for the shafts
shown in Figure 7.45 for horizontal alignment.

   5. Determine the appropriate sign (positive or negative) for
the side readings by applying the same rule in Step 3. For the
shaft arrangement shown in Figure 7.45, the appropriate signs
for the side readings are illustrated in Figure 7.49.
   6. Algebraically subtract the side readings from the bottom
reading and divide by two.

   DR to DN   DN to DR

   $\dfrac{(\pm B) - (\pm D)}{2}$ $\dfrac{(\pm A) - (\pm C)}{2}$

For the shaft arrangement shown in Figure 7.45 the values would
be:

   DR to DN   DN to DR

   $\dfrac{(+B) - (+D)}{2}$ $\dfrac{(-A) - (+C)}{2}$

   7. Add the values found in Step 6 to both sides and multiply
all readings by two to determine the desired cold alignment readings, as illustrated in Figure 7.50.

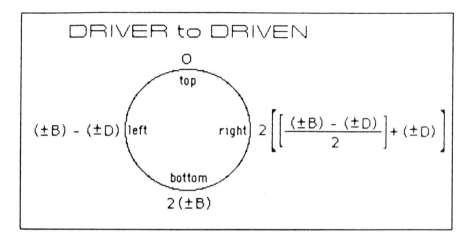

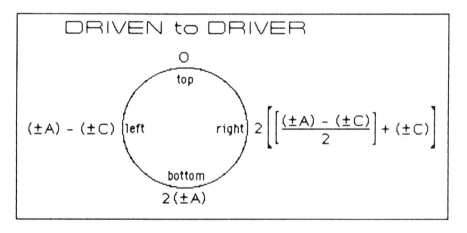

FIGURE 7.50 Equations for determining the "cold" shoot for readings.

For the shaft arrangement shown in Figure 7.45, the final desired cold reverse indicator readings are shown in Figure 7.51.

DRIVER to DRIVEN   DRIVEN to DRIVER

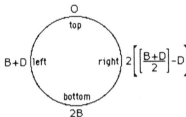
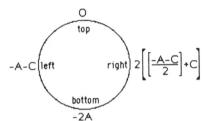

FIGURE 7.51 "Cold" reverse indicator readings for shafts shown in Figure 7.45.

Sample Problem

Figure 7.52 shows the desired cold shaft positions of a motor and blower arrangement based on field dynamic movements taken with a laser alignment system. The location of the quad cell detector targets is shown as triangles on the graph centerline, and the recorded movements are shown in Table 7.3.

Through the graphical plot of the desired cold shaft positions, Table 7.4 shows the gaps a dial indicator would see at the indicator reading points on the couplings of the motor and blower.

The corresponding desired cold reverse indicator readings are shown in Figures 7.53 and 7.54.

Sample Problem - Complete Alignment Solution

Thermal movement data for Figure 7.55 were taken using both the inside micrometer/inclinometer and optical tooling methods, with the results shown in Table 7.5. The averaged movement data are transferred to the graphical chart (Fig. 7.56) to obtain the "shoot-for" dial indicator readings and the shaft gapping information to determine the correct cold shaft relationships.

To assist in future calculations, the gap relationship between the actual shaft centerlines are marked at each foot, indicating

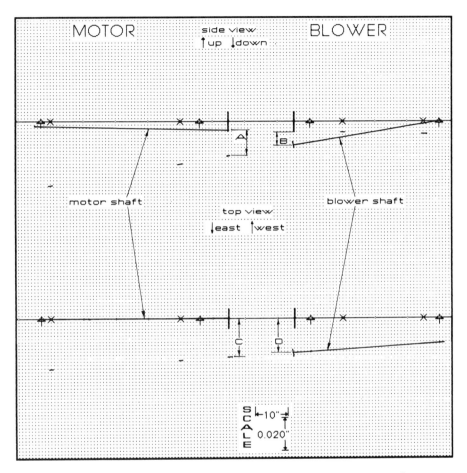

FIGURE 7.52 Plotted centerlines to show thermal movement from "cold" to "hot."

the amount and direction of the appropriate placement of the shaft. The side-to-side dial indicator readings are shown in the smaller circles. This helps in figuring the correct side readings to insure adherence to the validity rule (i.e., the sum of the two side readings equals the bottom reading).

TABLE 7.3 Movement Data of Motor and Pump

| | Vertical | | | | Horizontal | | | |
| | Outboard | | Inboard | | Outboard | | Inboard | |
	Amount	Direction	Amount	Direction	Amount	Direction	Amount	Direction
Motor	0.003 in.	Up	0.005 in.	Up	0	–	0	–
Blower	0	–	0.013 in.	Up	0.020 in.	East	0.015 in.	West

TABLE 7.4 Distances between Shafts Where the Dial Indicators Would Be Mounted

Bracket setup	Shaft view	Indicator location	Gaps	Position of indicator
Blower to motor	Vertical	A	-14	Bottom
Motor to blower	Vertical	B	+ 7.5	Bottom
Blower to motor	Horizontal	C	-23	East side
Motor to blower	Horizontal	D	+20.5	East side

MOTOR to BLOWER

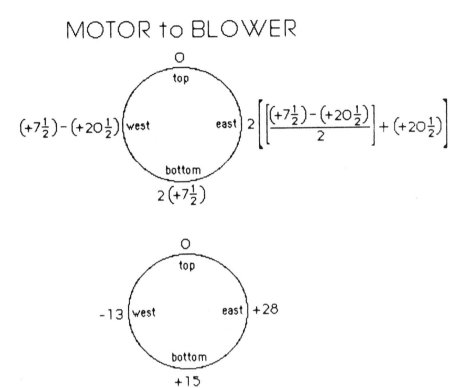

FIGURE 7.53 "Shoot for" cold readings on periphery of blower
shaft.

    Also shown on this graph setup are shaft "gapping notes"
that show the amount and direction between the actual shaft centerline and the projected shaft centerline of the other unit. These
are used as references when plotting the field readings onto the
chart. For instance, the compressor shaft should be placed 40
mils lower than the projected steam turbine centerline at the outboard foot and 12 mils lower than the projected steam turbine
centerline at the inboard foot, as shown in the vertical shaft position plot. Likewise, the steam turbine shaft can be referenced
to the compressor shaft in a similar manner (16 mils lower at the
inboard foot and 60 mils lower at the outboard foot). As the

BLOWER to MOTOR

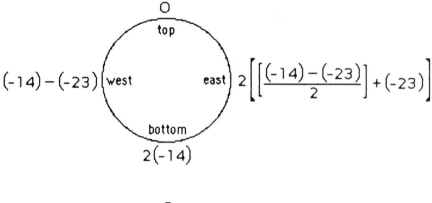

$$(-14)-(-23) \quad\quad\quad\quad\quad 2\left[\left[\frac{(-14)-(-23)}{2}\right]+(-23)\right]$$

$$2(-14)$$

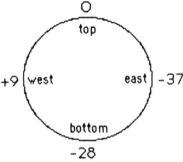

FIGURE 7.54   "Shoot for" cold readings on periphery of motor shaft.

field data are plotted, these "gap notes" will be shown on each alignment plot to insure that the correct cold shaft positions are maintained when calculating the required machinery moves.

With both units in place and bolted down, the first set of reverse indicator readings are taken, and the shafts are plotted in Figure 7.57. The shafts are considerably out of alignment sideways. The units will be positioned laterally first to increase the accuracy of the bottom readings. Since moving only one of the two units sideways would require a large lateral move at the outboard foot, a straight line overlay is projected from the

Compressor   Steam Turbine

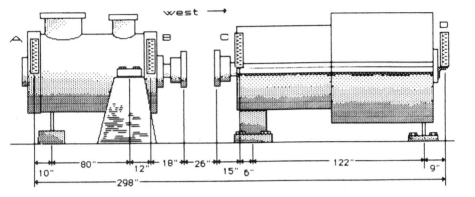

FIGURE 7.55 Compressor and steam turbine layout.

desired outboard foot location of the compressor to the desired
outboard foot location of the steam turbine. The inboard foot of
the compressor should then be moved to the south 115 mils and
the inboard foot of the steam turbine is moved 127 mils to the
south, as shown in Figure 7.57. If the steam turbine alone were
moved and the compressor remained stationary, the resulting
moves would have been 145 mils to the north at the inboard foot
and 400 mils to the north at the outboard foot, possibly resulting in the steam turbine becoming bolt bound by such large
moves.

Another alternative lateral move is shown in Figure 7.58 with
the straight line overlay in another position. After studying
these or other lateral movement alternatives by using the straight
line overlay technique, the optimum sideways move can be determined for effectively positioning the compressor and steam turbine before beginning the shim changes. Remember that the
shafts should not be directly in line since some movement occurs
when the units are brought up to operating speed, as observed
from the thermal movement calculations. Refer back to the thermal plots and use the "gapping notes" to calculate the moves
properly.

TABLE 7.5 Thermal Movement of Machinery from "Cold" to "Hot"

Inside micrometer/bevel protractor

Thermal measurement location	A		B		C		D	
Amount and direction	Amount	Direction	Amount	Direction	Amount	Direction	Amount	Direction
Horizontal	0.005 in.	North	0.002 in.	South	0	-	0.008 in.	North
Vertical	0.020 in.	Up	0.006 in.	Up	0.010 in.	Up	0.035 in.	Up

Optical Tooling

Thermal measurement location	A		B		C		D	
Amount and direction	Amount	Direction	Amount	Direction	Amount	Direction	Amount	Direction
Horizontal	0.003 in.	North	0	-	0	-	0.010 in.	North
Vertical	0.025 in.	Up	0.005 in.	Up	0.012 in.	Up	0.042 in.	Up

Averaged data comparing the two methods

Thermal measurement location	A		B		C		D	
Amount and direction	Amount	Direction	Amount	Direction	Amount	Direction	Amount	Direction
Horizontal	0.004 in.	North	0	-	0	-	0.009 in.	North
Vertical	0.022 in.	Up	0.006 in.	Up	0.011 in.	Up	0.039 in.	Up

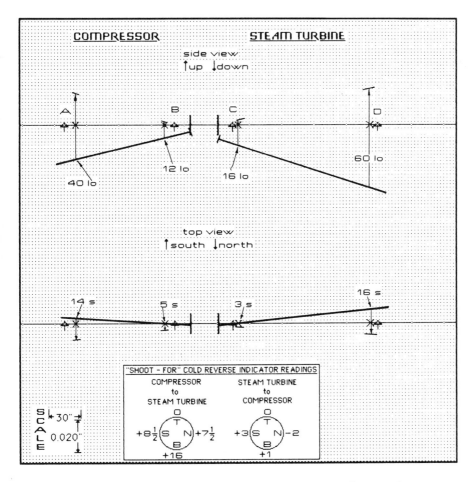

FIGURE 7.56 Plotted shaft centerlines and "shoot for" cold reverse indicator readings from thermal movement data.

After the lateral moves have been completed, another set of
reverse indicator readings are taken to begin properly positioning the compressor and steam turbine vertically. The readings
and graphical plot are shown in Figure 7.59.

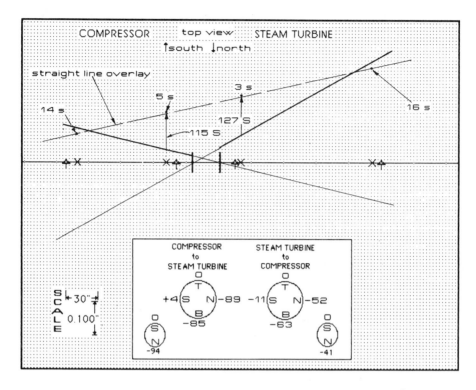

FIGURE 7.57 Plan view showing required sideways move of both units.

Once again the straight line overlay is used from one outboard foot to the other keeping in mind the proper shaft relations desired (Fig. 7.59). Here both inboard feet are raised by adding 138 mils to the inboard foot of the compressor and 168 mils to the inboard foot of the steam turbine. After the shims are added and the foot bolts torqued to the required value, a final set of reverse indicator readings yield the results shown in Figure 7.60.

The reverse indicator readings are close to the desired "shoot for" readings shown in Figure 7.56. Should another trim adjustment be made, or is this good enough for operating

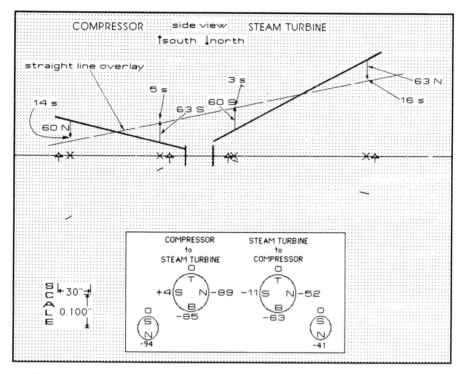

FIGURE 7.58 Alternative sideways move.

speeds at 12,500 rpm? Chapter 9 will shed some light on this question but the bottom line is, does the equipment run smoothly and trouble free for long periods of time? Chapter 10 will help in determining whether the alignment moves were effective or not.

BIBLIOGRAPHY

Alignment/Auto Collimating Laser System 71-2615 – Operating Manual. Manual no. 71-1000-4, Keuffel and Esser Co., Morristown, N.J., 1982.

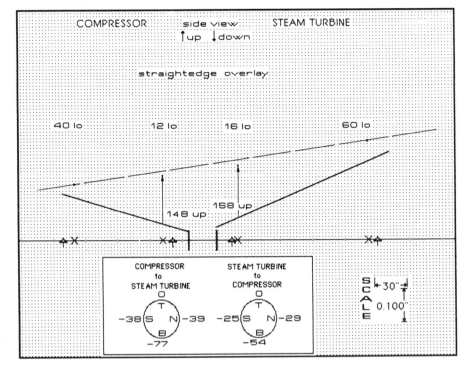

FIGURE 7.59 Vertical alignment plot showing required shim changes.

Applied Infrared Photography. Publication no. M-28, Eastman
    Kodak Co., Rochester, N.Y., May 1981.

Barnes, E. F. "Optical Alignment Case Histories." *Hydrocarbon
    Processing* (Jan. 1971), pp. 80-82.

Baumann, Nelson P. and William E. Tipping, Jr. "Vibration Re-
    duction Techniques for High-Speed Rotating Equipment."
    ASME paper no. 65-WA/PWR-3, Aug. 1965.

Blubaugh, R. L. and H. J. Watts. "Aligning Rotating Equipment."
    Chemical Engineering Progress (April 1969), pp. 44-46.

Baumeister, Theodore, Eugene A. Avallone, and Theodore Bau-
    meister, III. *Mark's Standard Handbook for Mechanical En-
    gineers*. 8th ed. McGraw-Hill, New York,

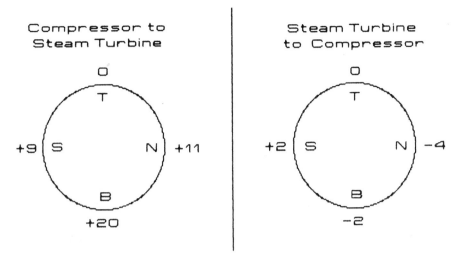

FIGURE 7.60 Reverse indicator readings after the sideways and vertical moves have been completed.

Campbell, A. J. "Optical Alignment Saves Equipment Downtime." *Oil and Gas Journal* (Nov. 1975).

Dodd, V. R. "Shaft Alignment Monitoring Cuts Costs." *Oil and Gas Journal* (Sept. 1972), pp. 91-96.

Dodd, V. R. *Total Alignment*. Petroleum Publishing Co., Tulsa, Ok., 1975.

Essinger, Jack N. "Hot Alignment Too Complicated?" *Hydrocarbon Processing* (Jan. 1974), pp. 99-101.

Hanold, John. "Application of Optical Equipment for Installing and Checking Large Machinery." ASME paper no. 61-PET-26, Aug. 1961.

Jackson, Charles. "Alignment with Proximity Probes." ASME paper no. 68-PET-25, Sept. 1968.

Jackson, Charles. "Successful Shaft - Hot Alignment." *Hydrocarbon Processing* (Jan. 1969), pp. 100-104.

Jackson, Charles. "How to Align Barrel-type Centrifugal Compressors." *Hydrocarbon Processing* (Sept. 1971), pp. 189-194.

Jackson, Charles. *The Practical Vibration Primer*. Gulf Publishing Co., Houston, Texas, 1979.

Koenig, Eugene. "Align Machinery by Optical Measurement." *Plant Engineering* (May 1964), pp. 140-143.

Lukacs, Nick. "Proximity Probe Applications for Troubleshooting Rotating Equipment Problems." I.S.A. paper no. 72-627, 1972.

Mitchell, John S. "Optical Alignment - An Onstream Method to Determine the Operating Misalignment of Turbomachinery Couplings." Dow Industrial Service, June 1972.

Mitchell, John W. "What is Optical Alignment?" Proceedings, Third Turbomachinery Symposium, Gas Turbine Labs, Texas A&M University, College Station, Texas, 1974, pp. 17-23.

Nelson, Carl A. "Orderly Steps Simplify Coupling Alignment." *Plant Engineering* (June 1967), pp. 176-178.

Norda, Torkel. "Use Infrared Scanning to Find Equipment Hot Spots." *Hydrocarbon Processing* (Jan. 1977), pp. 109-110.

O'Kelley, J. F. "Optical Shaft Elevation Measuring." *Power Engineering* (Oct. 1969), pp. 36-37.

"Optical Alignment - A Maintenance Service to Reduce Your Machinery Downtime." Bulletin no. 371-5000GP, Dresser Machinery Group, Dresser Industries, Houston, Texas, 1969.

Optical Alignment Manual. No. 71-1000, Keuffel and Esser Co., Morristown, N.J., 1969.

"The K&E Optical Leveling Kit and How to Use It." Bulletin no. T66-91222-66CT-3, Keuffel and Esser Co., Morristown, N.J., 1976.

Van Laningham, Fred L. "Distortion of Speed Changer Housings and Resulting Gear Failures." Proceedings, Fifth Turbomachinery Symposium. Gas Turbine Labs, Texas A&M University, College Station, Texas, Oct. 1976, pp. 7-13.

Yarbrough, C.T. "Shaft Alignment Analysis Prevents Shaft and Bearing Failures." *Westinghouse Engineer* (May 1966), pp.

8
Moving Machinery in the Field

Being able to calculate accurately the amount and direction that rotating equipment must be moved to bring the shafts into line may end up becoming an exercise in frustration if the machinery does not move the way it is supposed to. There is a monumental difference in knowing what the required amount of movement should be and being able to accomplish that desired movement. Positioning rotating machinery is an art that combines brute force with finesse, two qualities that usually are difficult to mesh together.

The steps for moving machinery may be summarized as follows:

1. Rough align the shafts in all three directions: axially, laterally, and vertically.
2. Position the units to obtain the proper shaft-to-shaft (axial) spacing.
3. Position the units sideways (laterally) as indicated from one of the calculation techniques.

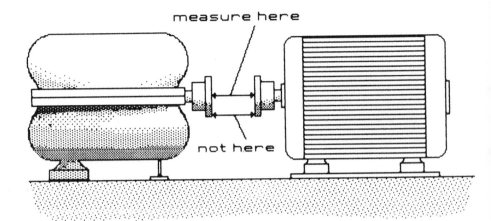

FIGURE 8.1 Measuring axial spacing.

4. Position the units vertically (shim change) as indicated from one of the calculation techniques.
5. Check axial spacing. Make trim adjustment if necessary.
6. Check lateral (sideways) position. Make trim adjustment if necessary.
7. Check vertical position. Make trim adjustments if necessary.
8. Repeat Steps 4 through 7 inclusive until desired accuracy is obtained.
9. Check and tighten the foundation bolts.

AXIAL SPACING

For many applications where rotors are supported by antifriction bearings (ball or roller type), this measurement is fairly straightforward. For rotors with hydrodynamic-type radial and thrust bearings, the actual position of the rotor under running conditions must be taken into account. Compressors, for instance, will run against the active thrust bearing surface and motors will run at their magnetic center.

The axial distance should be measured as close to the center-line of rotation of each shaft as possible (Fig. 8.1). If no toler-ance is given by the coupling manufacturer, the rule of thumb is to hold this distance to +0.010 in. of the recommended dimension. The importance of maintaining this gap cannot be overstressed, since coupling lock-up conditions occur as often from improper gap distances as from excessive vertical or lateral misalignment. The purpose of the coupling is to transmit rotational force, not thrust forces from one unit to another.

Many shafts may experience a change in length (i.e., "grow" longer or shorter) due to temperature rises or drops when sub-jected to operating processes. This change in axial dimension must be taken into account in the coupling and the machinery. Refer to Chapters 3 and 10 for further details on proper axial spacing on couplings, and what happens in the axial direction from a vibration standpoint.

LATERAL (SIDEWAYS) MOVEMENT

Jackscrews or whatever devices that may be used to slide equip-ment sideways should be placed as close as possible to the foot points without interfering with tightening or loosening of foun-dation bolts. A typical jackscrew arrangement is shown in Fig-ure 8.9. Dial indicators mounted on the baseplate that are used to monitor sideways movement should usually be placed on the opposite side of the machine case from where the movement de-vice (e.g., jackscrew) is located to keep the indicator from being bumped inadvertently.

Use a corner foot bolt as a pivot point and move one end of a unit at a time when moving sideways (Fig. 8.2).

After the outboard end has been moved the entire amount, tighten one of the outboard bolts and loosen the inboard pivot point bolt, as shown in Figure 8.3.

Monitor the movement of the inboard end either by placing a dial indicator at the side of the machine casing at the inboard foot or by using a dial indicator and bracket arrangement at-tached to one shaft, zeroing the indicator on one side of the coupling hub, then rotating the dial indicator and bracket ar-rangement 180° and noting the reading. Start moving the in-board end in the appropriate direction until the dial indicator

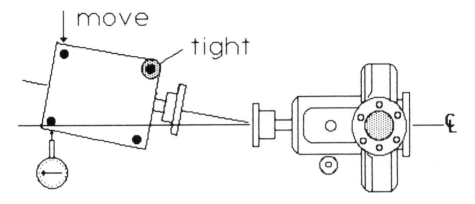

FIGURE 8.2 Slide the outboard end first.

on the coupling hub reads one-half the original value. Zero the
indicator again and rotate the dial indicator and shaft assembly
back 180° to the original zeroing point on the other side of the
coupling hub and check the reading on the indicator. If necessary, continue moving the inboard end to get the dial indicator
to read zero when swinging from side to side on the coupling hub

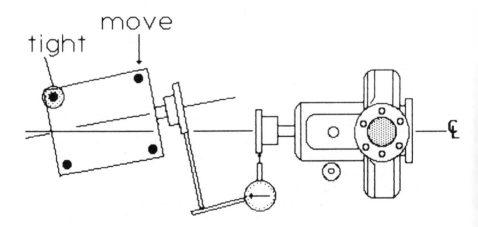

FIGURE 8.3 Move the inboard end by pivoting on one of the
outboard foot bolts.

TOTAL ALLOWABLE MOVEMENT MAP

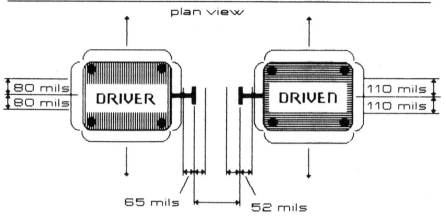

FIGURE 8.4 Total allowable sideways and axial movement.

(assuming there is no dynamic lateral movement in the machinery).
It is also possible to move both ends simultaneously using indicators and jackscrews at each foot. For equipment with inboard to
outboard foot distances of 3 ft or less, this seems to work well
with two people on the alignment job. Larger equipment usually
requires four or more people to be effective.

On new installations it may be desirable to generate a total
sideways movement "map." This will come in handy when calculating the necessary lateral moves to determine whether it is possible to move it as far as the calculation requires. An example of
a typical movement map is illustrated in Figure 8.4.

Once the map has been established, place each unit in the
center of its allowable sideways travel and begin to take readings at this point. These maps will prove invaluable when aligning multiple machine drive trains.

VERTICAL MOVEMENT

Lifting equipment is markedly more difficult than sliding it sideways, so it is desirable to make the minimum number of moves

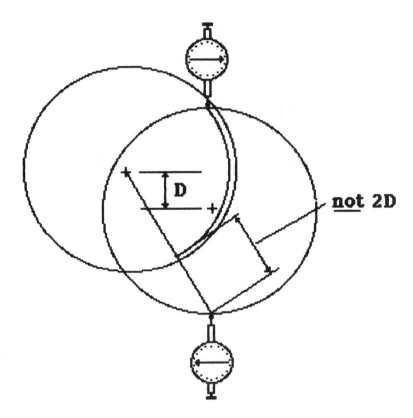

FIGURE 8.5 Bottom reading gives false vertical displacement of shafts.

necessary to achieve the correct vertical position. It is recommended that the units be aligned well sideways first before any
shims (besides what was needed for the rough alignment) are
added or removed from the feet. The reasoning for this is illustrated in Figure 8.5. Pretend you are viewing two shafts from
an axial vantage point and are able to observe the path of travel
of a dial indicator attached to one shaft and reading the circumference of the other shaft. Note the actual vertical displacement
with respect to both shaft centerlines. However, when the dial
indicator travels from the top to the bottom of the coupling hub,

FIGURE 8.6 Properly cut shim plates that "saddle" the foot bolts.

note that the indicator stem has traversed a distance different than 2D.

This reading inaccuracy will cause unnecessary shim changes later in the alignment process.

Shims or plates should be fabricated to properly saddle a foot bolt so the machine casting is not damaged when the bolt is tightened down. An example of this is illustrated in Figure 8.6.

If good lateral alignment has been achieved, try to keep as many foot bolts tight or have the jackscrews tightened against the machine element to prevent the unit from moving back out of alignment when shims are being added or removed from the feet. Lifting equipment with a couple of foot bolts tightened can be a very delicate and challenging operation and must be performed with extreme caution. The idea is to lift the unit just far enough to slide shims in or out.

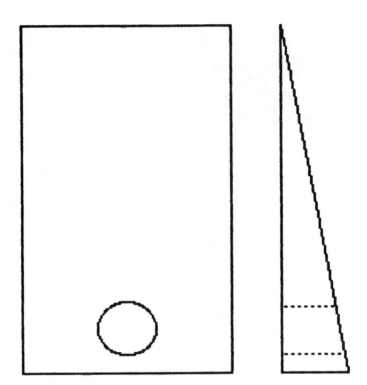

FIGURE 8.7 Hole drilled in wedge for removal.

TYPES OF MOVEMENT TOOLS

Hammers and Wedges

Hammers are probably the most widely used tool for moving machinery sideways, even if they are the least desirable. There
are preferred techniques for using hammers to move a unit sideways.

1. Use soft-faced hammers (plastic or rubber) instead of
steel hammers.

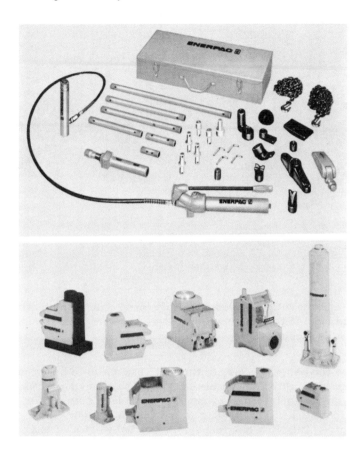

FIGURE 8.8 Hydraulic jacks. Courtesy of Enerpac, Butler, Wis.

2. If soft-faced hammers are not available, place a piece of wood or plastic between the hammer and the impact point on the piece of machinery to prevent damaging the case.

3. Take easy swings at first, then begin increasing the force. With practice, you can develop a feel for how the unit moves at each impact. The more force that is used, however, the greater chance there is to jolt the dial indicators that are monitoring the units movement rendering the readings useless.

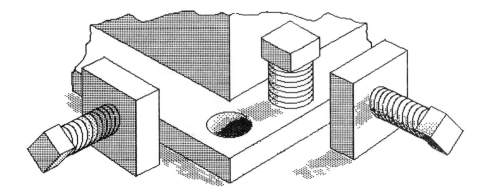

FIGURE 8.9 Typical jackscrew arrangement.

Using hammers and steel wedges to lift equipment is the least
desirable method. If there is no alternative, here are some tips
when using this technique.

1. Place the wedge as close to the foot that needs to be lifted
without interfering with the process of adding or removing shims.
The casing may distort enough to get the necessary shims in or
out of that foot area without having to lift the entire unit.
2. Apply a thin film of grease or oil to both sides of the
wedge.
3. It is fairly easy to install a wedge but it is quite another
thing to get it out from under a heavy piece of machinery. Provide some means of removal of the wedge, such as shown in Figure 8.7.

Pry and Crow Bars

Like sledge hammers, pry and crow bars can be found in every
mechanic's tool box, and they invariably end up at the alignment
site just in case they are needed. Consequently, for smaller,
light equipment, pry bars are the most widely used, yet least
desirable, device for lifting equipment. A pry bar provides very

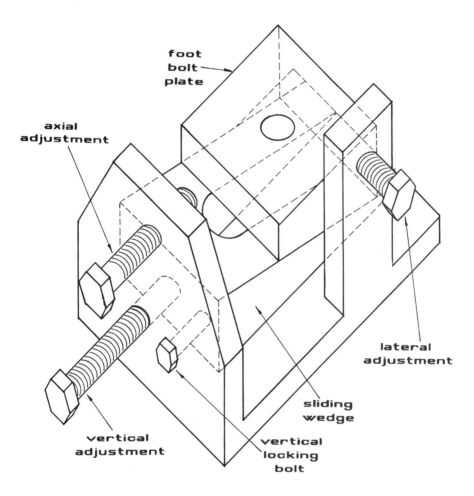

FIGURE 8.10 Permanent jackscrew and foot bolt device.

little control in lifting equipment accurately and can slip from its
position easily.

A pry bar can also be used to move the equipment sideways,
assuming there is a leverage device near the feet. The leverage
device, however, is usually piping, electrical conduit, or a long
piece of 2x4 supported against something else on the foundation.

FIGURE 8.11 Portable jackscrew for moving a 9000 hp motor.

Comealong and Chainfalls

These devices can be used both to lift and move equipment laterally. The primary problem with this equipment is usually the lack
of proper rigging or anchor points for the chainfalls or comealongs when moving sideways. There is also the problem of

FIGURE 8.12 Jackscrew used to move motor sideways.

exceeding the capacity of the chainfall when rigged to lift the
equipment. The better quality chainfalls and comealongs, however, provide more control and safety than do hammers and pry
bars.

Hydraulic Jacks

There are many types of hydraulic jacks and kits that can be purchased at reasonable prices (Fig. 8.8). When rigged properly,
hydraulic jacks provide good control and safety for lifting or
sliding equipment and are one of the preferred methods for moving
rotating machinery.

FIGURE 8.13 Machinery positioners. Photos courtesy of Murray & Garig Tool Works, Baytown, Tx.

Permanent Jackscrew Arrangements

Although jackscrews are undoubtedly the preferred method for moving machinery, they are not often used in industry, mainly because of the cost and effort required to install them. It is indeed a shame that all rotating machinery manufacturers and users cannot seem to find the time to provide these for their machinery.

A typical jackscrew arrangement at the corner of a machine element is illustrated in Figure 8.9.

If additional height is no problem, an arrangement similar to Figure 8.10 may be used on the equipment. These can be used

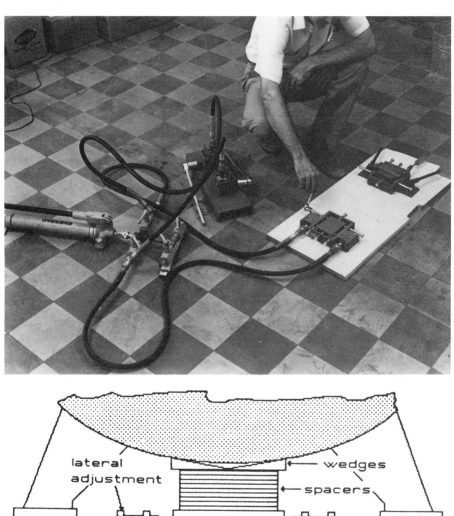

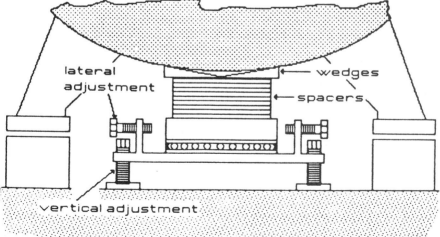

FIGURE 8.13 (continued)

to move the machinery while it is running, although you take a great risk in trying this.

Portable Jackscrews and Machinery Positioners

A considerable amount of imagination has gone into designing these clever devices and could be used in one form or another by most of industry for machinery alignment applications. If you have more than one of the same type of pump, motor, or compressor, it is recommended that devices such as shown in Figures 8.11 through 8.13 be used for your specific application.

DO'S AND DON'TS OF MOVING MACHINERY

DO'S

Place a thin film of lubricant under the foot points to facilitate lateral moves.

Needlenose pliers (not fingers) should be used to add or remove shims under the feet.

Consider removing shims instead of continually adding them under the feet.

Fabricate single-thickness plates instead of stacking many thin shims together.

Place shims under one side or foot at a time when making vertical moves to prevent the machine from losing its sideways alignment.

Radius the outside corners of thin shim stock to prevent them from folding over during installation.

When tightening the foundation bolts, have dial indicators positioned along the sides of each unit to determine if the unit is moving back out of alignment.

Apply the same amount of torque to the foundation bolts after each vertical or lateral move before taking another set of alignment readings.

Check the torque on all foundation bolts after completing the alignment job.

FIGURE 8.14 Cutting the shank of a bolt to allow for lateral movement.

DON'TS

Use pipe wrenches to rotate shafts or move equipment.
Use coupling bolts slid into the coupling hub holes and a pry bar
    to rotate a shaft for alignment readings.
Use carbon steel shim stock.
Use a hammer to install shims under the machinery feet.
Install shims that have paint anywhere on the surface of the
    shim.
Use a large number of thin shims to make a thick spacer under
    a foot.
"Chicago" foundation bolts (i.e., undercut the shank of the bolt),
    as shown in Figure 8.14.

Dowel your equipment. If the foot bolts cannot hold it down, dowels certainly will not. Refer to Chapter 7 on thermal movement to explain how machine casings expand and contract.

Attach the ground lead of an arc welding machine to turbomachinery casings or shafts when installing jackscrew blocks to the baseplate. Gear or similar metal-to-metal type couplings may fuse weld together at the transmission points.

NOTE: Always pick the smallest mechanic to rotate the heaviest rotors and move the biggest equipment...let the big, strong guys cut the shims and watch the indicators!

BIBLIOGRAPHY

Machinery Alignment Positioners. Murray and Garig Tool Works, Baytown, Texas, Dec. 1978.

Mannasmith, James and John D. Piotrowski. "Machinery Alignment Methods and Applications." Vibration Institute meeting, Cincinnati Chapter, Sept. 1983.

Massey, John R. "Installation of Large Rotating Equipment Systems - A Contractors Comments," Proceedings, Fifth Turbomachinery Symposium, Gas Turbine Labs, Texas A&M University, College Station, Texas, Oct. 1976.

Piotrowski, John D. "The Graphical Shaft Alignment Calculator." *Machinery Vibration Monitoring and Analysis*, Vibration Institute, Clarendon Hills, Ill., 1980.

9
Alignment Tolerances

Undoubtedly the most important question to be answered is: How accurate must shaft alignment between equipment be?

The preceding chapters have explained what is involved from a hardware standpoint and how to align the rotating equipment. Now we must address what is trying to be achieved.

Minimize amount of wear in coupling components.
Eliminate possibility of shaft failure from cyclic fatigue.
Minimize the amount of shaft bending from the point of power transmission in the coupling to the coupling end bearing.
Minimize the force on the bearings to insure long bearing life and rotor stability under dynamic operating conditions.
Minimize vibration in machine casings, bearing housings, and rotors.

Protection and longevity of each component in the drive train is the key goal.

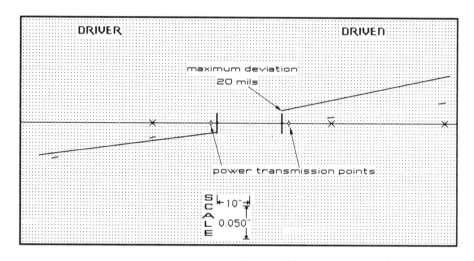

FIGURE 9.1 One mil per inch misalignment condition.

Misalignment is the deviation of relative shaft position from a colinear axis of rotation, measured at the points of power transmission when equipment is running at normal operating conditions.

For instance, if the points of power transmission are 20 in. apart and the maximum shaft centerline to projected shaft centerline offset is 20 mils, the deviation is one mil per in. of power transmission distance. At 1800 rpm, this deviation is acceptable; at 20,000 rpm, the alignment deviation is unacceptable. The alignment plot shown in Figure 9.1 illustrates a one mil per inch maximum misalignment condition.

A general guideline for alignment tolerances is shown in Figure 9.2. Acceptable amounts of misalignment must be tailored to suit each individual drive train application. As mentioned previously, gear-type couplings and universal joint drives must have slight amounts of misalignment for proper lubrication to occur. Staying within the *acceptable* misalignment band usually fills this requirement. Diaphragm couplings, on the other hand, should be aligned within the *excellent* range. Frequently, well-aligned equipment may pose problems with lightly loaded shafts in plain

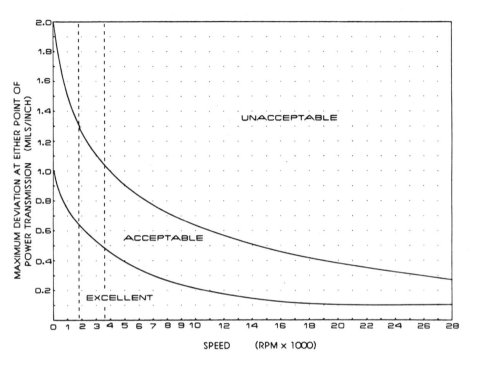

FIGURE 9.2 Guidelines for alignment tolerances.

hydrodynamic bearings, causing a whirl phenomena. In cases like this, position the shafts to have the maximum misalignment deviation occur near the bearing experiencing the whirl. Misalign the equipment only enough to prevent instability. One contractor has deliberately used electric heaters on the support legs of a compressor to misalign a unit while running. Shaft orbit information from X-Y proximity probes monitored the vibration response until stability was attained at that bearing. By recording the change in temperature and calculating the change in support leg height, shims were then added to stabilize the whirl. Although it is not recommended, changing the alignment of machinery while it is running has successfully been done by a few people.

BIBLIOGRAPHY

Dodd, V. R. *Total Alignment*. Petroleum Publishing Co., Tulsa, Ok., 1975.

Piotrowski, John D. "The Graphical Shaft Alignment Calculator." *Machinery Vibration Monitoring and Analysis*. Vibration Institute, Clarendon Hills, Ill., 1980.

10
Finding Misalignment on Rotating Machinery

Undoubtedly, the single most important parameter in determining the health of rotating machinery is the level of vibration that exists in each of the drive train elements. The performance of the equipment relating to its designed operating conditions, such as output horsepower, discharge pressure, flow, speed, etc., may all be just fine, but if excessive vibration of the equipment is present, it is definitely not going to be running for very long. For some people, frequent overhauls of machinery seem to be taken as a fact of life, and many organizations routinely rebuild equipment whether it is needed or not. Worse yet, some let their machinery run itself into the ground.

Since the late 1960s, a radically different philosophy has emerged. Preventive maintenance programs are being merged with predictive maintenance techniques that have taken the operation of rotating equipment to a higher plateau of improved performance by attempting to predict the exact time of failure as it steadily increases in vibration. Of course, this does not always prevent the instantaneous failures that will occur, but it will stop machinery from slowly beating itself into oblivion. This requires a commitment by management, engineering, maintenance, and operating personnel to monitor operating parameters more closely, collect and interpret vibration data, and understand how all this relates to the overall performance of the equipment. Today, it is almost taboo not to install some sort of vibration monitoring system on new machinery installations. Many existing systems have been retrofitted with vibration sensors and monitors to watch the overall vibration levels of rotors and bearings to prevent many failures. For smaller, less expensive rotating machinery, handheld vibration meters are used and data are taken at regular intervals. This vibration data is trended over periods of time until one or more sensors indicate that a level is beginning to approach, attain, or exceed a predetermined high limit and it is time to correct the problem. But what is the problem?

Discerning what a vibration sensor is telling you happens to be one of the most difficult tasks facing the machinery diagnostician. However, with the use of a fast fourier transform (FFT) signal analyzer, vibration "signatures" can be taken that split the complex overall vibration signal and enable one to look at various frequencies of the sensor's output. Many rules of thumb have emerged that attempt to classify specific machinery problems with specific types of vibration signatures. The experienced vibration analyst quickly learns that these rules of thumb are to be used as guidelines on arriving at the source of the problem. Quite often, more than one problem exists on a piece of rotating machinery, such as a combination of imbalance, misalignment, and damaged bearings that will all appear on the vibration spectrum.

The purpose of this chapter is to examine the types of vibration signatures misaligned rotating equipment exhibit and the forcing mechanisms involved to generate this signal.

CONDUCTING A THOROUGH ANALYSIS

In addition to using vibration analysis as a tool to determine a misalignment symptom, the field engineer should also make additional sensory checks (eyes, ears, touch) of the suspect machine and reflect on some of the following items.

Equipment Inspection

Are the bearings hot to the touch? (Threshold of pain is around
   120°F. If your hand doesn't like it, neither will the bearing!)
Is there an excessive amount of oil leakage at the bearing seals?
Do the foundation bolts loosen up periodically? Are they loose
   now?
Is there cracking in the grouting or concrete on the foundation?
Are the coupling bolts loose? Are there any bolts missing; do
   they show wear or excessive stresses from cyclic fatigue or
   overtorquing?
Are the coupling bolts part of a matched set?
If the coupling is grease packed or a nonlubricated type, is it
   hot immediately after shutdown? If it is an elastomeric type,
   is there rubber powder inside the coupling shroud?
Does the shaft runout have a tendency to increase after opera-
   ting the equipment for some time?
Does similar equipment seem to be vibrating less or have a longer
   operating life?
Is there a high number of failures in the coupling or a problem
   with shafts breaking at or close to the inboard bearings?
Is there grease on the inside of the coupling guard?

TAKING VIBRATION DATA

Vibration data should be taken at five points on each machine
element, as shown in Figure 10.1. If possible, a once per rev
signal should be taken and phase angle data be recorded, particularly in the axial direction. Be careful when taking phase
readings at the coupling end of each unit, since the sensor will
probably be pointing in opposite directions in the axial direction, which produces a 180° phase shift.
   If overall vibration amplitude levels are relatively high at
any of the sensor locations, a more detailed survey should be
taken with a spectrum (FFT) analyzer to aid in determining the
vibration signature of the machinery. If the source of the problem is misalignment, what will the vibration spectrum look like?
Does the signal look the same for all different types of machinery
with different kinds of couplings, bearings, and rotors?

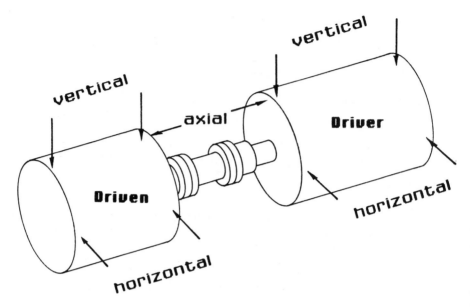

FIGURE 10.1   Vibration sensor locations.

HOW VARYING DEGREES OF MISALIGNMENT
AFFECT ROTATING MACHINERY –
A CASE STUDY

The pump and motor arrangement as shown in Figure 10.2 was
used to study what effect different kinds of shaft alignment have
on the vibration of this equipment.

Equipment Data

Motor: 60 hp; 1775 rpm; cast aluminum construction; rotation is
     clockwise when standing at the outboard end of motor looking
     at pump.

FIGURE 10.2 Pump and motor used for test.

Pump: size, 8 X 10-12; 2000 gpm; 85-ft head.
Coupling: metallic ribbon type; alignment limit, 10 mils (angular and offset); Figure 10.3 shows the coupling design.

    Thermal movement of the pump and motor arrangement was calculated by taking surface temperature measurements at various points on the pump and motor, as shown in Figure 10.4.
    The temperatures shown were taken after the unit was operating for two hours with good shaft alignment, allowing for thermal stability to be achieved. The thermal movement on the motor was very slight, with the inboard end raising 5 mils and the outboard end raising only 2 mils. Since no change in temperature occurred on the pump from off line to running conditions, there was no change in position of the pump shaft centerline.

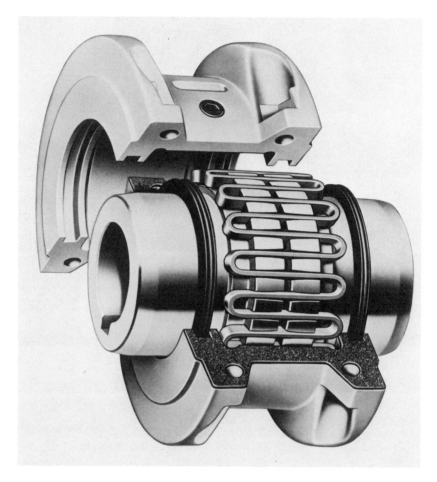

FIGURE 10.3 Coupling design. Courtesy of Falk Corp., Milwaukee, Wis.

It is important to note that the surface temperatures at various points on the motor, particularly the inboard end bell housing of the motor, increased as the amount of misalignment increased. The maximum temperature observed occurred as one of the extreme misalignment conditions attained 145°F. Each test run lasted approximately 30 min, so the motor did not have a chance to stabilize thermally.

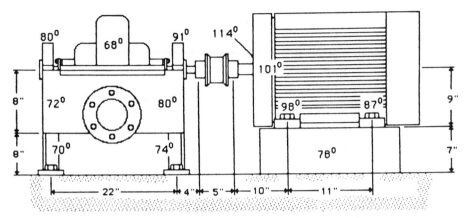

FIGURE 10.4 Surface temperatures of pump and motor when running (all temperatures in degrees Fahrenheit).

Alignment Test Runs

A total of seven test runs were made under various alignment conditions. Vibration data were taken at five points on each unit using a handheld vibration meter with an accelerometer sensor to record overall readings on the bearing housings. A seismometer with a magnetic base was also attached at each sensor location, and the signal was fed into a analyzer and an X-Y plotter to record the vibration signature. Since the motor housing was aluminum construction, 1/4-in.-thick carbon steel plates were epoxied to the motor in the horizontal and vertical planes at both bearings, and another plate was attached on the inboard end bell to capture the axial vibration levels. The vibration monitoring equipment is shown in Figure 10.5.

Shaft Alignment Positions During the Test Runs

Run No. 1 (Motor uncoupled and running solo)

The first run was conducted with the motor uncoupled to determine whether the motor had an imbalance condition, damaged

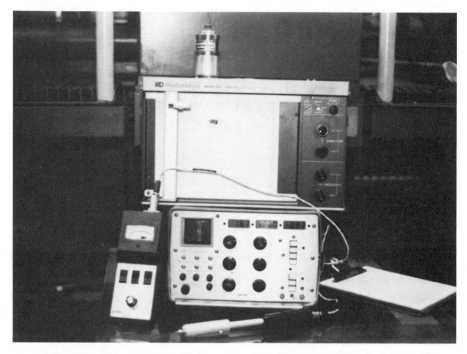

FIGURE 10.5 Vibration monitoring equipment. Photo courtesy of
Scientific Atlanta, Atlanta, Ga., and IRD Mechanalysis, Columbus, Oh.

bearings, or some other problem that could affect the vibration
response when coupled to the pump.

Run No. 2. M2W (Motor-2 mils-west)

Figure 10.6 shows the shaft position of the pump and motor with
good alignment conditions. The reverse-indicator shaft alignment method was used to determine the relative shaft positions
for all test runs. The nomenclature used to identify each run
indicates the maximum amount of motor shaft deviation with
respect to the pump shaft at the point of grid contact in the
coupling. In this run, for instance, the maximum amount of shaft

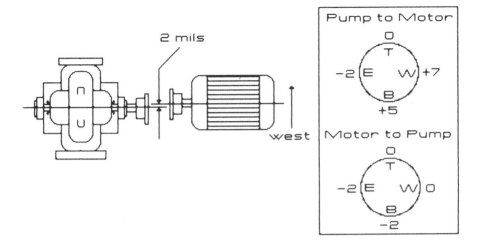

FIGURE 10.6 Motor-2 mils-west.

deviation occurred in the lateral (i.e., side-to-side) position of
the shafts. The motor coupling hub contact point is 0.002 in. to
the west of the contact point on the pump coupling hub and is
designated as M2W or motor-2 mils-west.

Run No. 3. M21W (Motor-21 mils-west)

Figure 10.7 shows the position of the motor shaft 21 mils to the
west of the pump shaft with virtually no deviation in the vertical position. In other words, the motor was positioned 0.021 in.
to the west with no shims being added or removed from the
pump or the motor from the second run.

Run No. 4. M36W (Motor-36 mils-west)

The attempt was made in this run to slide the motor further to
the west. The motor however became bolt bound and was unable to be moved any further sideways, as shown in Figure
10.8.

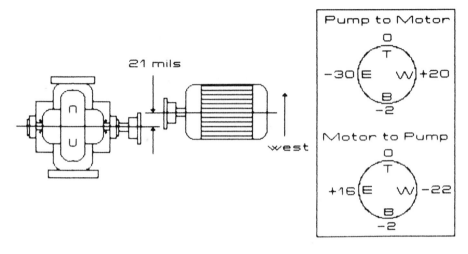

FIGURE 10.7 Motor-21 mils-west.

Run No. 5. M65H (Motor-65 mils-high)

The motor is now positioned well from side to side but is 0.065 in.
higher than the pump shaft centerline, as shown in Figure 10.9.

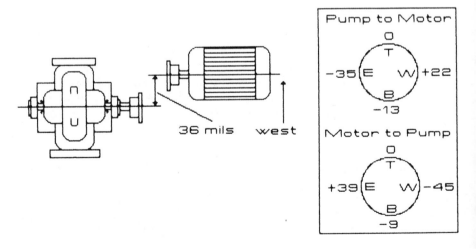

FIGURE 10.8 Motor-36 mils-west.

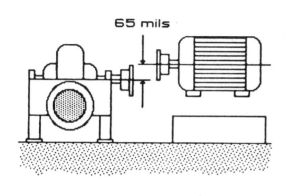

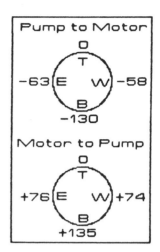

FIGURE 10.9 Motor-65 mils-high.

Run No. 6.   M55L (Motor-55 mils-low)

Figure 10.10 shows that the motor shaft centerline is pitched below the pump shaft centerline while still maintaining good side-to-
side alignment.

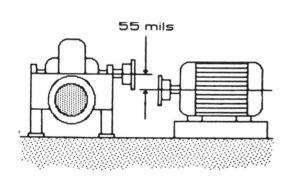

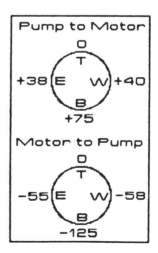

FIGURE 10.10 Motor-55 mils-low.

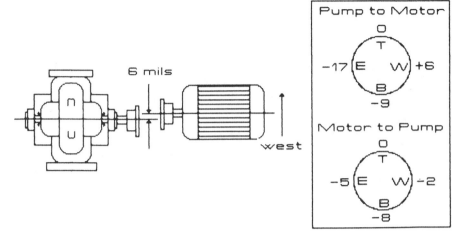

FIGURE 10.11 Motor-6 mils-west.

TABLE 10.1 Overall Vibration Amplitude Levels on Bearing Housings

			Overall vibration levels[a]					
	Run no.	Motor solo	M 2W	M 21W	M 36W	M 65H	M 55L	M 6W
Motor	North							
	Horizontal	0.03	0.06	0.07	0.07	0.12	0.13	0.07
	Vertical	0.02	0.04	0.06	0.07	0.12	0.10	0.05
	South							
	Horizontal	0.02	0.05	0.04	0.07	0.17	0.19	0.04
	Vertical	0.02	0.04	0.04	0.10	0.21	0.14	0.05
	Axial	0.02	0.04	0.04	0.08	0.15	0.12	0.05
Pump	North							
	Horizontal	–	0.09	0.05	0.15	0.21	0.24	0.10
	Vertical	–	0.10	0.04	0.07	0.10	0.12	0.06
	Axial	–	0.09	0.05	0.12	0.19	0.23	0.10
	South							
	Horizontal	–	0.10	0.06	0.13	0.15	0.19	0.11
	Vertical	–	0.04	0.03	0.05	0.09	0.10	0.01

[a]All readings in in./sec.

Run No. 7. M6W (Motor-6 mils-west)

As shown in Figure 10.11, the motor was repositioned within acceptable alignment tolerance levels (similar to Run 2, M2W) to determine if the vibration response at the bearings would repeat
what was found in Run 2.

Vibration Data

The overall vibration amplitude levels measured on the bearing
housings with the handheld vibration meter are shown in Table
10.1. The vibration spectrum plots are shown in Figures 10.12

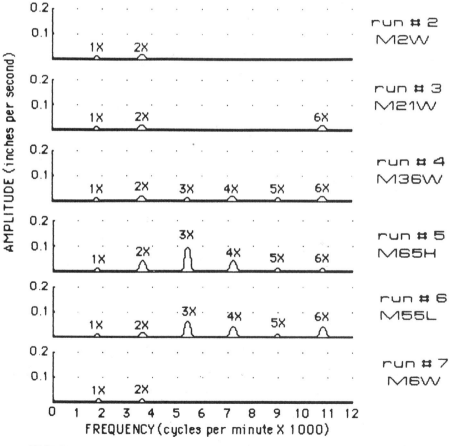

FIGURE 10.12 Outboard motor, horizontal direction.

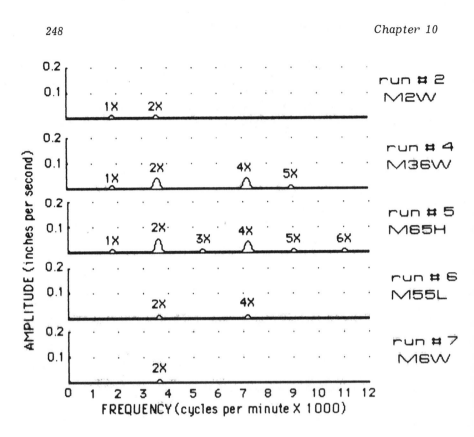

FIGURE 10.13 Outboard motor, vertical direction.

through 10.22 and were arranged to observe the vibration signature changes during each test run at each of the sensor locations
on the motor and the pump.

Observations and Comments – Overall Vibration Levels

The largest change in overall vibration levels occurred at the inboard bearings of both the pump and the motor.

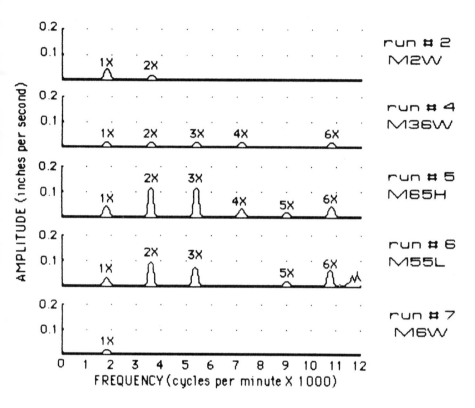

FIGURE 10.14 Inboard motor, horizontal direction.

The axial vibration level on the motor attained its highest value
    when its shaft was higher than (above) the pump shaft. Like-
    wise, the axial vibration level on the pump attained its highest
    value when its shaft was higher than the motor shaft.
The highest horizontal amplitude readings on both the pump and
    the motor occurred when the motor shaft was low with respect
    to the pump shaft centerline (M55L), even though the amount
    of misalignment was not as severe as in the M65H case.
The pump bearings increased in amplitude when the motor shaft
    was higher than the pump shaft (from Run 5 to 6), whereas
    the motor bearings decreased or stayed the same.

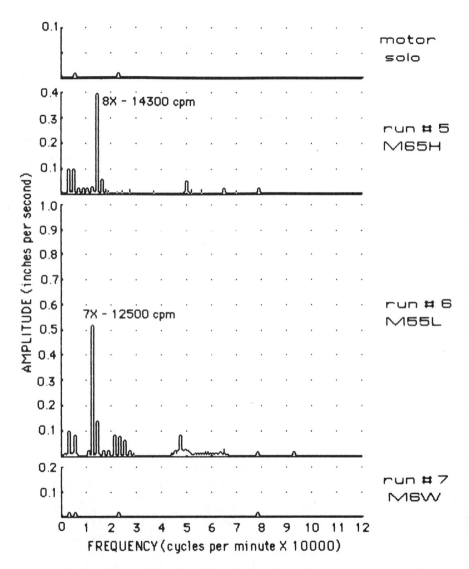

FIGURE 10.15 Inboard motor, horizontal direction (0-120,000 cpm).

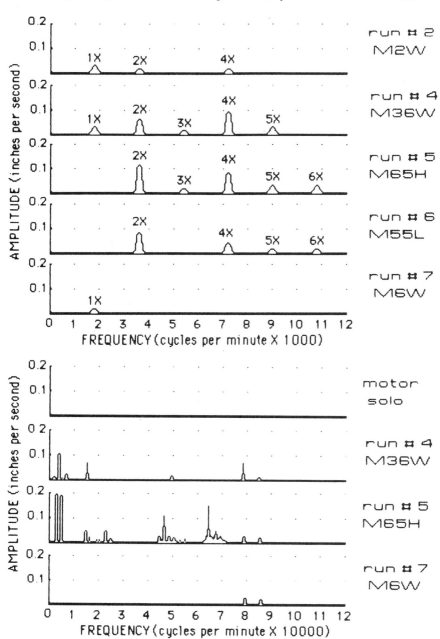

FIGURE 10.16 Inboard motor, vertical direction.

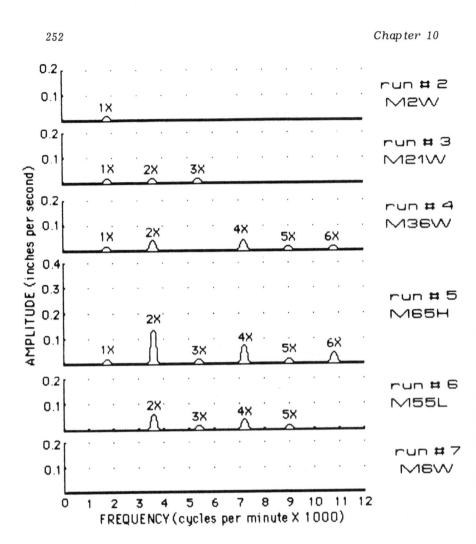

FIGURE 10.17 Inboard motor, axial direction.

The outboard bearing of the pump experienced a greater increase
in overall vibration levels compared to the outboard bearing
of the motor.

The pump experienced a decrease in overall vibration levels on
both bearings when the motor was misaligned laterally (Runs
2 and 3).

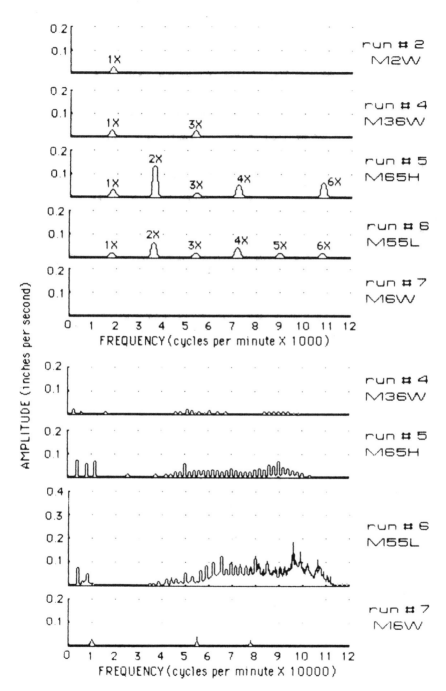

FIGURE 10.18 Inboard pump, horizontal direction.

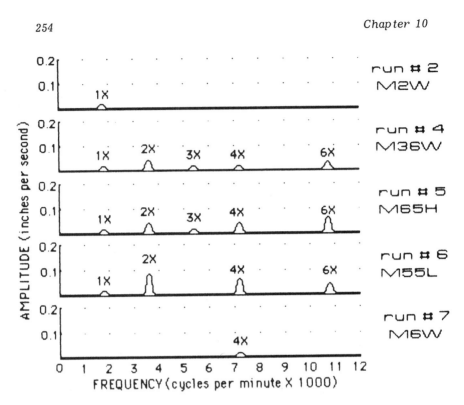

FIGURE 10.19 Inboard pump, vertical direction.

The horizontal and axial amplitude levels of the inboard pump
bearing increased as the amount of the misalignment increased. The vertical levels, however, changed only slightly
during all the different shaft configurations.

The overall vibration levels on the pump and motor taken during
Run 7 were, for the most part, the same as the levels taken
during Run 2, where the alignment conditions were nearly
identical, verifying that the vibration was due to misalignment and not other factors.

Observations and Comments – Vibration Spectrums

The once per rev amplitude levels were not affected by any
misalignment conditions and, in fact, decreased in many

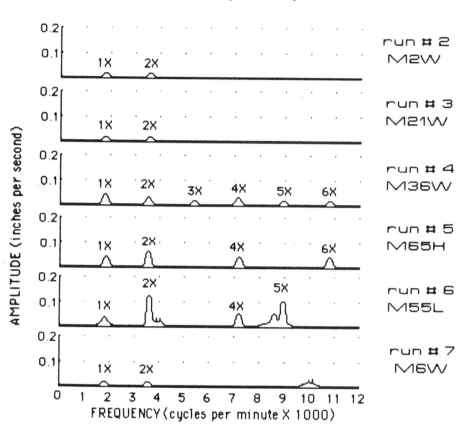

FIGURE 10.20 Inboard pump, axial direction.

instances when the misalignment increased. Consequently, the attempts at taking phase-angle data when tracking the once per rev signal were inconclusive, because the phase angle would continually drift.

The highest peaks occurred at the maximum misalignment conditions (M65H and M55L) on the inboard sensor locations on the motor.

The 3X peaks on the motor were higher in amplitude in the horizontal directions than in the vertical direction.

The 2X and 4X peaks on the motor were more dominant in the vertical and axial directions than in the horizontal direction.

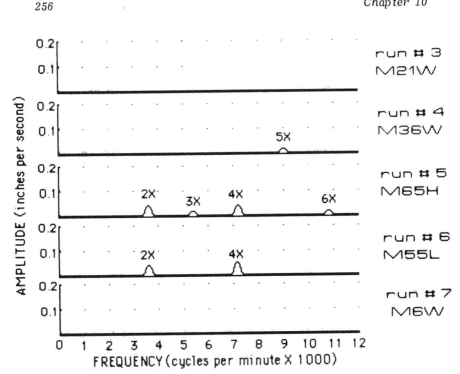

FIGURE 10.21 Outboard pump, horizontal direction.

The 2X, 4X, and 6X peaks prevailed in the pump bearings. Higher multiples of running speed occurred on the pump from 40 to 100 kcpm, particularly during the vertical misalignment runs.

Analysis of Vibration Data

The twice, fourth, and eighth running speed frequencies are a result of the "S" shaped grid as it traverses from its maximum tilted and pivoted positions twice each revolution on both the coupling hubs, as shown in Figure 10.23.

The third and sixth running speed multiples occur as the metal grid in the coupling changes its position during each revolution of the shafts. The maximum amount of rotational force

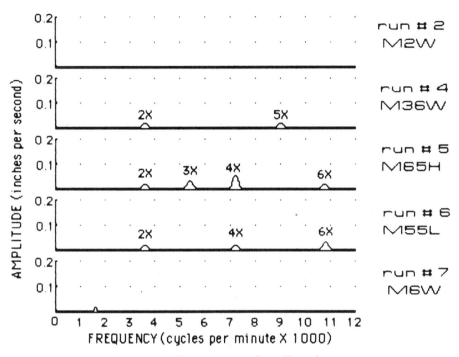

FIGURE 10.22 Outboard pump, vertical direction.

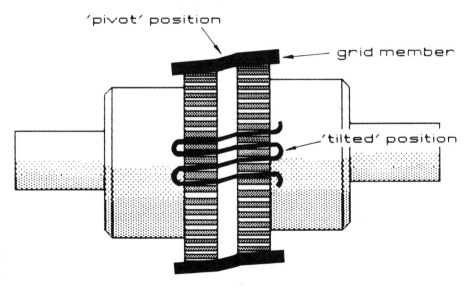

FIGURE 10.23 Position of metal coupling grid under misaligned conditions.

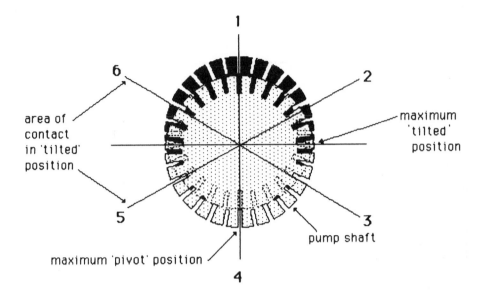

FIGURE 10.24 Axial view of both shafts when misaligned.

occurs when the grid is in the tilted position where bending occurs across the thickness of the grid member. Imagine being
able to look axially down both the motor and pump shafts, as
shown in Figure 10.24. From point 1 to point 2, the "S" shaped
grid is undergoing a transition from a "pivoted" position to a
"tilted" position, where a minimum amount of rotational force is
being applied to the shafts.

As the grid enters the area near point 2, the rotational force
begins increasing until it passes through the maximum tilted position where the applied torque peaks, and then subsides as it
continues through to point 3. From point 3 to point 4, the force
continues to diminish as the transition from tilt to pivot occurs,
but now in a direction opposite from the point 1 pivot condition.
As the grid continues on through points 5, 6, and back to 1,
the sequence repeats itself but in a reversed direction.

The position of the shafts in Figure 10.24 closely resembles
conditions M65H or M55L where there is no lateral misalignment
and considerable vertical misalignment. Points 2, 3, 5, and 6

do not actually occur at 60° arcs, and as the vertical alignment improves, the area of contact in the tilted position increases and the rotational force is more evenly distributed across the entire grid. The seven times running speed peak that occurred in the horizontal direction on the inboard motor bearing during the M55L run, and the five times running speed peak that occurred in the axial direction on the pump are, as yet, not completely understood as to the source of the forcing mechanism involved. The higher multiples appear to be caused by overloading the anti-friction bearings.

Conclusions

The vibration signatures of the pump and motor displayed multiples of running speed, with the predominant frequencies occurring at 2X, 3X, 4X, 5X, 6X, 7X, and 8X (i.e., higher order harmonics). The once per rev amplitude levels were not affected by any of the various shaft alignment positions regardless of the severity of shaft misalignment. The higher amplitudes of vibration occurred at the inboard bearings of both the pump and motor. The dynamic action that occurs in the coupling with this particular design is the determining factor in the vibration signature of the equipment.

OTHER MISALIGNMENT VIBRATION
SIGNATURES

An aircraft derivative gas/power turbine driver set coupled to two axial flow compressors via a gear coupling with an 80-in. span between shaft ends was experiencing a significant increase in vibration levels when accelerated from idling conditions to full-load operating speed. Figure 10.25 shows the vibration signature taken at the coupling end bearing of the power turbine. When idling, the predominant peaks were occurring at once per rev on the gas turbine (6800 cpm), twice per rev on the power turbine (9300 cpm), and four per rev on the power turbine (18600 cpm).

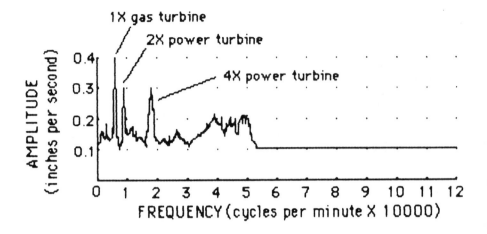

FIGURE 10.25 Power turbine vibration at coupling end bearing at idle speed.

When aligned "cold," the gas/power turbine shaft centerline was set 40 mils to the east of the compressor centerline. With the 80-in. span between the points of power transmission in the gear coupling, the resulting misalignment was 1/2 mil per in.

Figure 10.26 shows the vibration signature when the driver set was accelerated to full-load operating speed (7400 rpm on the gas turbine, with a resulting speed of 5100 rpm on the power turbine).

Even though the increase in speed was slight, the change in the vibration spectrum was quite dramatic. The predominant peak occurring at running speed on the gas turbine and the four times running speed component on the power turbine increased substantially.

Water-cooled proximity probe stands were mounted at the gas turbine inlet and at the coupling end bearing of the power turbine to monitor the movement of the driver set. At idle conditions, the power turbine shaft centerline dropped 15 mils. When accelerated to full-load operating speed, the coupling end of the gas/power turbine then moved 55 mils to the east. With the centerline of rotation of the driver set already 40 mils to the east when aligned "cold," the additional 55 mils of movement further

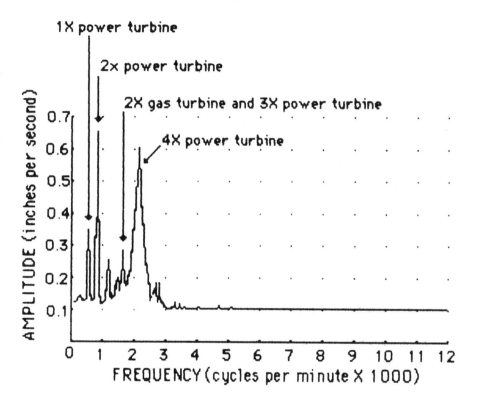

FIGURE 10.26 Power turbine vibration at coupling end bearing at full-load operating speed.

to the east produced slightly over 1 mil per in. misalignment, which is unacceptable at this operating speed.

Figures 10.27 and 10.28 are vibration signatures taken on two different 9000-hp induction motors. Data on Figure 10.27 were recorded with a proximity probe monitoring the shaft motion.

Figure 10.28 was taken with a seismometer (velocity pickup) mounted on the bearing housing. Notice how the three times running speed frequency shows up on the vertical and axial direction, and the twice running speed frequency in the horizontal direction. The misalignment for both these two different motors exceeded 2 mils per in.

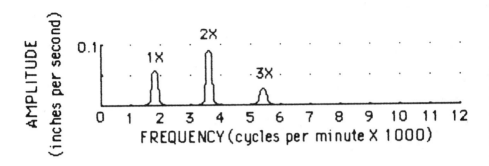

FIGURE 10.27 Vibration signature of a 9000 hp motor with a proximity probe sensor monitoring the horizontal shaft vibration.

In each of the misaligned drive trains shown in this chapter, there is a recurring pattern in the vibration spectrums that exhibits multiples of running speed (harmonics). The amplitudes of each of these components will change under varying alignment conditions.

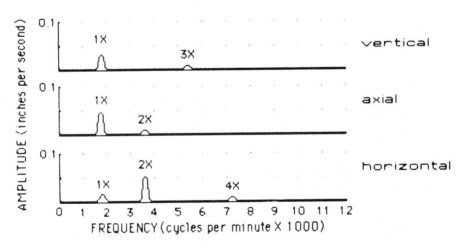

FIGURE 10.28 Vibration of motor bearing at the coupling end.

FIGURE 10.29 Failure of flexible disc coupling after two years service on misaligned cooling tower drive system.

It must be pointed out, however, that this data only covers two coupling designs: flexible grid and gear-type couplings. There is still much more investigation needed for other types of couplings on different types of machinery at various operating conditions.

BIBLIOGRAPHY

"Alignment Loading of Gear Type Couplings." Application notes no. (009)L0048. Bently Nevada Corp., Minden, Nevada, March 1978.

Audio Visual Customer Training - Instruction Manual. Publication no. 414E. IRD Mechanalysis Inc., Columbus, Ohio, 1975.

Baxter, Nelson L. *Vibration and Balance Problems in Fossil Plants: Industry Case Histories.* Publication no. CS-2725. Research project no. 1266-27, Electric Power Research Institute, Palo Alto, Ca., Nov. 1982.

Bertin, C. D. and Mark W. Buehler. "Typical Vibration Signatures - Case Studies." *Turbomachinery International* (Oct. 1983), pp. 15-21.

Eshleman, Ronald L. "Torsional Vibration of Machine Systems." Proceedings, Sixth Turbomachinery Symposium, Gas Turbine Labs, Texas A&M University, College Station, Texas, Dec. 1977.

Eshleman, Ronald L. "Effects of Misalignment on Machinery Vibrations." Proceedings, Balancing/Alignment of Rotating Machinery, Galveston, Texas, Feb. 1982.

Eshleman, Ronald L. "The Role of Couplings in the Vibration of Machine Systems." Vibration Institute meeting, Cincinnati Chapter, Nov. 1983.

Jackson, Charles. *The Practical Vibration Primer.* Gulf Publishing Co., Houston, Texas, 1979.

Mannasmith, James and John D. Piotrowski. "Machinery Alignment Methods and Applications." Vibration Institute meeting, Cincinnati Chapter, Sept. 1983.

Maxwell, J. H. "Vibration Analysis Pinpoints Coupling Problems." *Hydrocarbon Processing* (Jan. 1980), pp. 95-98.

Piotrowski, John D. "How Varying Degrees of Misalignment Affect Rotating Machinery - A Case Study." Proceedings, Machinery Vibration Monitoring and Analysis Meeting, New Orleans, La., June, 1984.

Piotrowski, John D. "Aligning Cooling Tower Drive Systems." Proceedings, Machinery Vibration Monitoring and Analysis. Ninth annual meeting, May, 1985, New Orleans, La.

Sohre, John S. "Turbomachinery Analysis and Protection." Proceedings, First Turbomachinery Symposium. Gas Turbine Labs, Texas A&M University, College Station, Texas, 1972.

Appendix

MACHINERY DATA CARD

Equipment Name:_____ **Location**:_____

Equipment Sketch – Side View (show overall dimensions)

DRIVER Information

☐ stm. turb. ☐ gas turb. ☐ ind motor ☐ sync. motor ☐ diesel ☐ other_____

MFG._____ Mdl. No._____ Ser. No._____
HP._____ RPM _____ Ser. Fact. _____ Trip speed _____
Critical Speeds_____ Total Weight _____ lbs. Rotor Wt. _____
Shaft Dia. _____ in.   Shaft taper _____ mils/in.
Interference fit _____ mils NO. of Keyways _____ Key size _____ x _____ x _____ in.
BEARINGS: ☐ antifriction ☐ sleeve ☐ other (type) _____
clearance _____ mils Rotor axial float _____ mils
type of lubricant _____ amount of lubricant _____

DYNAMIC MOVEMENT: outboard _____ mils up down _____ mils N S E W (circle directions of motion)

inboard _____ mils up down _____ mils N S E W (circle directions of motion)

Foot bolt size_____ in.   wrench size _____ in.

DRIVEN Information

☐ compressor ☐ pump ☐ gear ☐ generator ☐ blower ☐ other _____

Mfg._____ Mdl. No._____ Ser. No._____
Size_____ Capacity _____
Other applicable information _____

Critical speeds _____ Total Weight _____ lbs. Rotor Weight _____ lbs.
Shaft dia. _____ in.   shaft taper _____ mils/in.
interference fit _____ mils No. of keyways _____ key size _____ x _____ x _____ in.

BEARINGS: ☐ antifriction ☐ sleeve ☐ other (type) _____
clearance _____ mils Rotor axial float _____ mils
type of lubricant _____ amount of lubricant _____

DYNAMIC MOVEMENT: outboard _____ mils up down _____ mils N S E W (circle directions of motion)

inboard _____ mils up down _____ mils N S E W (circle directions of motion)

Foot bolt size_____ in.   wrench size _____ in.

COUPLING Information

☐ gear ☐ diaphragm ☐ elastomeric ☐ chain ☐ flex disc ☐ flex link

☐ leaf spring ☐ metal ribbon ☐ pin drive ☐ other _____

Mfg._____ Mdl./Size _____ Ser. No _____

Coupling bolt size: _____ Bolt Torque _____ in/lbs Wrench Size _____

Shaft to Shaft spacing_____ in. +/- _____ in.

Type of lubrication _____ Amount of lube _____

SHAFT TO SHAFT ALIGNMENT INFORMATION   (be sure to indicate the <u>direction</u> of the side readings)

Desired 'cold' alignment readings

DRIVER to DRIVEN

O

T

B

DRIVEN to DRIVER

O

T

B

'COLD' SHAFT POSITION SKETCH

Vertical Position

Horizontal Position

MISCELLANEOUS INFORMATION

Date unit installed: _____ - _____ - _____

Tools needed to move equipment for alignment: ☐ jackscrews   wrench size _____ in.

☐ hydraulic jacks (tonnage) _____   ☐ other tools needed _____

Other Information _____

_____

ALIGNMENT CHECKLIST

Equipment

Is the equipment indoors or outdoors? (circle one)

When can the equipment be shut down to do the alignment?
_____

If the equipment is down now, when does the job have to be completed? _____

How many shifts will be working during the course of the day to carry out the alignment? _____

What form of communication should be used to inform each shift of the progress being made? _____

What type of equipment is being aligned? _____

If the equipment is a compressor or a pump, will the piping be attached during the alignment? □ yes □ no

If the driver is an electric motor, who is responsible for lowering the breaker and safety tagging the switchgear?
_____ (name)

If the driver is a steam turbine, who is responsible for shutting off the steam supply and safety tagging the valve?
_____ (name)

Does the coupling spool need to be removed? □ yes □ no

Should the coupling bolt set be replaced due □ yes □ no
to damage?

Where can the coupling pieces be stored so they don't get lost or damaged? _____

Has the coupling been properly match marked for reinstallation?
□ yes □ no

What type of grease does the coupling take, how is it applied, and how much grease is needed? _____ type
_____ amount
□ grease gun □ hand packed □ other _____

If the coupling is a continuous oil feed system, can the supply lines be plugged or capped so the lube system can be run to assist in rotating the shafts for alignment readings? □ yes □ no

Will the coupling hubs have to be removed from the shafts?
☐ yes ☐ no

What is the coupling bolt torque? _____ in.'lbs.

How can the shafts be rotated without damage?
_____

Can the shafts be rotated in both directions? ☐ yes ☐ no

How heavy are the rotors? _____ lbs (driver)
_____ lbs (driven)

How heavy is the equipment? _____ lbs (driver)
_____ lbs (driven)

Do the shafts rotate freely?   ☐ yes   ☐ no

Is there lubricant on the bearings?   ☐ yes   ☐ no

What are the shaft diameters that have to be turned?
_____ in. (driver)
_____ in. (driven)

What alignment method will be used? ☐ reverse-indicator ☐ face-peripheral ☐ other _____ (circle one)

What are the dimensions of the equipment? (*see* Machinery Data Card)

Can the foundation bolts be loosened easily?   ☐ yes   ☐ no

Is the environment conducive for work or is there noise, heat, cold, vibration, etc. that might hamper work progress? ☐ yes ☐ no

If the equipment is to be removed for overhaul, get a set of alignment readings before removal and use a centerpunch on the machine casing at its base and the foundation to assist in quickly relocating the equipment when it is installed. ☐ completed ☐ yet to be done ☐ does not apply

Is the foundation and baseplate sound, or will repairs have to be made? ☐ yes ☐ no _____ repairs needed

What size are the foundation bolts and how many are there?
            bolt size (in.)   qty.   wrench size (in.)
driver _____   _____   _____
driven _____   _____   _____

Are there shims under the foot points now? ☐ yes ☐ no

How thick are the shims under all the feet?

	inboard (mils)	outboard (mils)
driver	_____	_____
driven	_____	_____

Could a lot of thin shims stacked together be replaced with a single plate? ☐ yes ☐ no

What type of material was used for the existing shims? ☐ carbon steel ☐ stainless steel ☐ brass ☐ other
_____

Is the baseplate clean and rust free? ☐ yes ☐ no

Are the feet of the equipment clean and rust free? ☐ yes ☐ no

What are the dynamic movement characteristics of the rotating equipment? (*see* Machinery Data Card)

Does the equipment have a soft foot? ☐ yes ☐ no
location _____
amt. _____ (mils)

What provisions are there for moving the equipment? ☐ jack-screws ☐ hydraulic jacks ☐ portable machinery movers ☐ other _____

Is there an excessive amount of runout on the coupling hub or shaft? ☐ yes ☐ no

Is the shaft marked for 90 degree turns for stopping the dial indicator as it traverses the coupling hub? ☐ yes ☐ no

Tools

What type of dial indicator bracket will be used?
_____

What are the bracket sag characteristics? _____
sag at bottom (mils)

What tools are needed to rotate the shafts?
_____

What tools are needed to attach the alignment bracket to the shaft?
_____

What size wrenches are needed for:
foundation bolts _____,
jackscrews _____ ,
coupling, etc.? _____

What is the stem travel of the dial indicators, what type are they,
are they reliable? _____ (in.) stem travel
_____ type

Will more than one dial indicator be available?   ☐ yes   ☐ no
_____ qty.

Are the magnetic bases for the dial indicators in good condition?
☐ yes   ☐ no

What tools or measuring devices will be used to measure the shaft
to shaft distances?   ☐ inside micrometers   ☐ taper/feeler gauges
☐ other _____

Where can the tools be stored overnight if needed?
_____

If no jackscrews are provided for moving the equipment sideways
or up or down, what will be used for substitution?
_____

Paper and pencil for recording readings, measurements, etc.

People

Who will supervise the overall job coordinating the engineering and
trades effort, collect and record measurements, readings, etc.?
_____

Who will rotate shafts, remove couplings, move the equipment, cut
shims, etc.? _____

Who will calculate the necessary moves for completion of the alignment and decide if the proper tolerances have been met?
_____

Assign specific duties for each person to perform.
_____
_____
_____

If two or more crews are working on the job, allocate a certain amount of overlap of shifts to communicate the progress made and the tasks yet to be completed.

Are these people experienced in alignment or is training neces-sary? ☐ yes ☐ no

Index